Implications of Climate Change and Disasters on Military Activities

NATO Science for Peace and Security Series

This Series presents the results of scientific meetings supported under the NATO Programme: Science for Peace and Security (SPS).

The NATO SPS Programme supports meetings in the following Key Priority areas: (1) Defence Against Terrorism; (2) Countering other Threats to Security and (3) NATO, Partner and Mediterranean Dialogue Country Priorities. The types of meetings supported are generally "Advanced Study Institutes" and "Advanced Research Workshops". The NATO SPS Series collects together the results of these meetings. The meetings are co-organized by scientists from NATO countries and scientists from NATO's "Partner" or "Mediterranean Dialogue" countries. The observations and recommendations made at the meetings, as well as the contents of the volumes in the Series, reflect those of participants and contributors only; they should not necessarily be regarded as reflecting NATO views or policy.

Advanced Study Institutes (ASI) are high-level tutorial courses to convey the latest developments in a subject to an advanced-level audience.

Advanced Research Workshops (ARW) are expert meetings where an intense but informal exchange of views at the frontiers of a subject aims at identifying directions for future action.

Following a transformation of the programme in 2006, the Series has been re-named and re-organised. Recent volumes on topics not related to security, which result from meetings supported under the programme earlier, may be found in the NATO Science Series.

The Series is published by IOS Press, Amsterdam, and Springer, Dordrecht, in conjunction with the NATO Emerging Security Challenges Division.

Sub-Series

A.	Chemistry and Biology	Springer
B.	Physics and Biophysics	Springer
C.	Environmental Security	Springer
D.	Information and Communication Security	IOS Press
E.	Human and Societal Dynamics	IOS Press

http://www.nato.int/science
http://www.springer.com
http://www.iospress.nl

Series C: Environmental Security

Implications of Climate Change and Disasters on Military Activities

Building Resiliency and Mitigating Vulnerability in the Balkan Region

edited by

Orlin Nikolov
CMDR COE, Sofia, Bulgaria

Swathi Veeravalli
Geospatial Research Laboratory, Engineer Research and Development Center
Vicksburg, Mississippi, USA

Springer

Published in Cooperation with NATO Emerging Security Challenges Division

Proceedings of the NATO Advanced Research Workshop on Implications of Climate Change and Disasters on Military Activities: Building Resiliency and Mitigating Vulnerability in the Balkan Region
Sofia, Bulgaria
5–7 July, 2016

Library of Congress Control Number: 2017939938

ISBN 978-94-024-1073-0 (PB)
ISBN 978-94-024-1070-9 (HB)
ISBN 978-94-024-1071-6 (eBook)
DOI 10.1007/978-94-024-1071-6

Published by Springer,
P.O. Box 17, 3300 AA Dordrecht, The Netherlands.

www.springer.com

Printed on acid-free paper

Preface

The primary concern of the current generation is the changes that are occurring in the global climate and environment. We all know that changes are inevitable, whilst nothing is permanent. The situation was entirely different hundred years ago, and reversals nowadays brought about many alterations in our attitude towards the climate. Obviously, the natural climate changes are beyond our control, and hereafter, these sheets are aiming to impacts occurring in earth-atmosphere-ocean system due to human interference.

The Crisis Management and Disaster Response Centre of Excellence (CMDR COE) supported by the NATO SPS Programme conducted an Advanced Research Workshop (ARW) on topic Climate Change Implications on Military Activities in the Balkan Region. The event took place from 5 to 7 July 2016 in Sofia.

The event was co-chaired by Col. Katarina Strbac (Serbian Ministry of Defence) and Col. Orlin Nikolov (CMDR COE). The AR Workshop was attended by participants from 17 nations – Albania, Bosnia and Herzegovina, Brazil, Bulgaria, Georgia, Greece, FYROM, Hungary, Montenegro, the Netherlands, Poland, Romania, Serbia, Turkey, Ukraine, the UK and the USA.

The expected outcome is to build a continuous capacity of partnership to enhance interoperability between local, national and regional institutions dealing with environmental security issues through developing an annual course focused on potential climate change impacts on NATO and partner nations in the region.

Generally, the aim of the ARW 2016 is to provide preliminary recommendations for education and capacity development in regard to the new security challenges posed by climate change. However, the present proceeding observes the changes reported and sees the grim climate change scenarios that are being projected by today's theoretical and mathematical models.

In order to perform the aforementioned goal, the subject matter experts involved in the ARW 2016 evaluated current and envisaged future capability needs by analysing security environment and identifying possible domains of civil-military cooperation in the climate change field in the Balkans and Southeastern Europe.

The editors gratefully acknowledge the contributions to the conclusions and recommendations by the participants. The ARW 2016 brought together lecturers from the US European Command, US Africa Command; US Army Corps of Engineers; US Army Command and General Staff College; Asia-Pacific Center for Security Studies; US Agency for International Development; International Alert; Tennessee Emergency Management Agency; Tennessee National Guard; Environmental Law Institute; Data for Resilience; National Institute for Meteorology and Hydrology, Bulgarian Academy of Sciences; Directorate General Fire Safety and Civil Protection, Ministry of Interior of the Republic of Bulgaria; and Hellenic National Defence General Staff.

We appreciate the willingness of the authors of the various chapters.

Special thanks are due to Ms. Swathi Veeravalli for the preparations of the event and conscientious handling of the large amount of correspondence involved and Prof. Chris King and Dr. Imes Chiu for their careful support during the entire workshop phases.

Sofia, Bulgaria — Orlin Nikolov
Vicksburg, MS, USA — Swathi Veeravalli

Acknowledgment

(Crisis Management and Disaster Response Centre of Excellence)

Chief Editor:

Orlin Nikolov – *Chief of Capabilities Branch*

Associate editors:

Jaroslaw Kubisz – *Chief of Transformation Branch*
Plamen Milanov – *Concept and Experimentation Section Head*
Panagiotis Aposporis – *Ddoctrine and Standardisation Section Head*
Milen Milkov – *Chief of Education and Ttraining Branch*

Managing editor:

Boris Guenov – *Publications, Library and Database Expert*

Technical support:

Desislav Zmeev – *Planning and Projects Section Head*
Svetlin Denchev – *Publications, Library and Database Expert*
Plamen Petrov – *CIS Specialist*
Gergana Vaklinova – *Doctrine and Standardisation Expert*

Address:
34A Totleben Boulevard, Shipka Hotel, floor 2
1606 Sofia, Bulgaria
Phones: 359 29224700, 359 2 9224705

Acknowledgments

Address

On the occasion of the workshop 'Implications of Climate Change and Disasters on Military Activities: Building Resilience and Mitigating Risks and Vulnerability in the Balkans'

5–7 July 2016

Sofia, Bulgaria

Dear generals and officers,
Dear distinguished guests,
Ladies and gentlemen,

It is an honour and a great pleasure to warmly welcome you at our workshop on the implications of climate change and disasters on military activities, the ultimate goal of which is to contribute to strengthening resilience on risk and vulnerability mitigation in the Balkan region. The Crisis Management and Disaster Response Centre of Excellence (CMDR COE) has the privilege to host this important event bringing together representatives, including keynote speakers and participants, from 18 countries which is a clear sign of the CMDR COE commitment and ambition to serve as a wide platform for information exchange and best-practices sharing thus enabling lasting partnerships on topical matters in the CMDR domain.

The CMDR COE, as an accredited international military organisation and a NATO military body, has as an objective to contribute to the enhancement of NATO's, NATO nations' and partners' capabilities on CMDR through research, analyses, lessons-learned and information sharing. To that end, the mission of the centre is to serve as a primary focal point on crisis management and disaster response.

Since its creation and bearing testimony to the ceaseless efforts of the centre and its staff members in creating and maintaining a community of interest in the CMDR domain, the centre has generated a wealth of information and expertise; forged excellent partnerships and enhanced cooperation with key NATO and EU bodies, institutions and agencies; and participated, hosted and organised an array of training courses (one NATO listed and four NATO selected), seminars and conferences in the aforementioned field.

In a global context of dynamic and highly volatile security environment, we are all faced by pressing challenges, increasing in both number and complexity, and multiplying threats which, respectively, require innovative collective responses based on strong leadership and commitment to peace and security. As a politico-military alliance, the NATO has stood the test of time, namely, because it stands prepared to embrace the aforementioned change in the strategic environment and to adapt its structure, functioning and objectives accordingly.

Ladies and Gentlemen,

The comprehensive character of the CMDR requires a clear understanding of and an attention to, inter alia, the changing security landscape as a result of climate change. Our home, our planet, our natural environment is changing, and at a pace never registered before, as a result of our activity or, and more importantly, of our inactivity. Climate change is a decisive global challenge which, as it remains unmanaged, endangers global stability and security.

The severity of natural disasters continues to increase causing unprecedented stress to the entire ecosystem on the planet ultimately threatening its long-term survival. Ladies and gentlemen, we are all witnessing, and experiencing, extreme weather conditions worldwide which condemn people to hunger (scarce resources) or to undertaking, sometimes life-threatening, mass movements in search of better living conditions, and even to death. For the sake of forsaking self-centrism, it should be noted that the adverse consequences of climate change are spread both horizontally across our species and vertically across the entire ecosystem.

Natural and man-made disasters become ever more complex to tackle, particularly given the simultaneity of occurrence, and interconnected, which necessitates joint, adaptive, well-planned and coordinated actions between main stakeholders at national, regional and international level, including in humanitarian assistance and disaster relief.

Ladies and Gentlemen,

Essentially, leadership, commitment and cooperation – these are the basic ingredients of any response that strives for effectiveness. And an effective operational response adopts the principles of the comprehensive approach (CA) to address multi-faceted conflicts and crises and to contribute to sustainable and lasting peace. Such an approach calls for coordination of efforts particularly in light of the widespread repercussions that both natural and man-made disasters have on our globalised, interconnected and interdependent world. Complementarity of efforts is thus a critical component in terms of effective mitigation.

Therefore, ladies and gentlemen, I would like to finally stress that climate change is not a construction but the reality we are living in. And we may not be able to reverse effects and processes already triggered, but we can take the responsibility to work towards creating the necessary conditions (structures and mechanisms) for minimising the negative consequences for future generations, for our children and grandchildren.

It is my strong conviction that we must finally 'walk the talk', but before that, we have to make sure we know our objectives and how to join efforts in achieving them.

I thank you for your attention and wish you all a productive, constructive and informative work.

Director of the Crisis Management
and Disaster Response Centre
of Excellence (CMDR COE)

By Col. Vassil Roussinov

Welcome Address

By Mrs. Zlatina Karova

Director of 'Science' Directorate, Ministry of Education and Science

NATO Science for Peace and Security Advanced Research Workshop 'Implications of Climate Change and Disasters on Military Activities: Building Resilience and Mitigating Risks and Vulnerability in the Balkans'

5 July 2016, 9 AM – Arena di Serdica Residence Hotel

Dear Colonel Roussinov,
Dear organisers,
Dear colleagues and guests,

Firstly, I would like to express my heartfelt gratitude for the opportunity to take part in today's workshop, in which will be discussed a number of issues of global concern related to the development of the modern world, including the climate change and its implications.

I clearly emphasise my full support for the mission of the NATO Crisis Management and Disaster Response Centre of Excellence, located in Sofia. Improving interoperability, developing and enhancing NATO standards in building capacities for crisis management and disaster response and providing assistance for the effective use of joint experience and capabilities of the NATO, member states, partners and international organisations in the field of crisis management and disaster response will contribute to the successful development of NATO policy, doctrine and concepts related to crisis management and disaster response activities.

This 3-day workshop and its further development as an annual course for understanding the potential impacts of climate change on military operations is fully compatible with the NATO Science for Peace and Security Programme, established in 1958.

The Ministry of Education and Science initiated the establishment of the interdepartmental working group, in order to develop a mechanism for interaction and cooperation with NATO.

Bulgaria actively participates in this programme since 1993. Our country's participation was most intense in the 1998–2000 period within the context of the government's decision to join the alliance. Over the past 20 years, over 230 scholarships have been granted, and over 600 Bulgarian scientists have been supported by the Science Fellowship subprogramme for research or lecturing in research institutions.

In 2015, the Ministry of Education and Science organised the national competition 'Young scientists with a prominent interest in the field of defense and security', and its aim was to find, motivate and promote the ideas of young and ambitious people, graduates or Ph.D. students with research interests in the field of defence and security, giving them the opportunity to realise their ideas with senior scholars from research groups in the Republic of Bulgaria and member states and partners of NATO. The candidates had to develop their own conceptual projects within the priority topics for Bulgaria and the region:

- Newly emerging challenges in security.
- Increasing the support for operations and missions led by NATO.
- Improving the awareness about the changes in the security environment, including through early warning in order to prevent crises.
- Humanitarian and social aspects, related to the strategic objectives of NATO.
- Any project that has a clear connection with security threats outside these priorities was considered for funding by the NATO Science for Peace and Security (SPS) Programme.

In 2015, the Ministry of Education and Science organised an Information Day for the NATO Science for Peace and Security Programme. The Information Day was attended by representatives of Serbia and Montenegro, and the discussion was about the possibilities for the development of regional cooperation and further interaction mechanisms for the implementation of joint projects.

In conclusion, I want to emphasise that the research of the implications of climate change-induced natural disasters on military activities in the Balkan region and identifying both policy and technical approaches to building resiliency and flexibility will reduce regional vulnerability to potential devastating disasters.

I would like to wish you all success, and I believe that our joint efforts will contribute to further defining the problems and challenges that climate change poses to civil-military missions and objectives in the Balkans and other disaster-prone areas in Eastern Europe!

Thank you for your attention!

Address

By Col. Ph.D. Katarina Strbac
Director of Directorate for European Integration and Project Management
Ministry of Defence, Republic of Serbia

Distinguished guests and dear participants, welcome to Advanced Research Workshop on Implications of Climate Change and Disasters on Military Activities.

The topic on the first sight is quite unusual, but recent history in the world and in our region has changed the archaic attitude that the military is invented to run wars and that security is exclusively a question of peace and war. Today's world is a much more complicated place than before. We have today changed system characteristics of our world: the world became more dependent on humans. The twentieth century will be known as a century of atom, cosmos, world wars, catastrophes and disasters. A new reality came to our eyes: the progress of science and development of technological sphere caused a number of serious challenges, risks and threats to humans and their environment. World development dynamics analysis recognises two key tendencies, each related to separate threats. The first tendency refers to a sudden development of human activity that changes the world. The second one might conditionally be called globalisation of the issue, consequently leading to a special 'globalisation of concern'. Strategy of survival and development of some countries and regions, at the account of others, has no historical perspective. If, in the past, we were only able to talk about a direct link between economic growth and expanded reproduction of national economies, we now have a bundle of complex relations among environment, agriculture, energy consumption, emergencies, countries' development strategies and international cooperation.

Just 10 years ago, national security was a synonym for defence, but today, national security has a much wider military meaning. The reason is the fact that modern threats are mostly asymmetric in nature, especially in the phase of development and threat/risk exploitation, when it is possible to act in a preventive way to stop the creation of conflicts and the escalation of crisis. Considering that the definition of security according to modern threats is very wide, scientific programmes with security as their core research area must be redefined to include topics and areas that explore those phenomena. Security has to contain military aspect as its core, but considering the changed nature of threats, it is required to provide a different military response that cannot only be passive defence or preventive measures. In the development and reaffirmation of security sciences, it is required to generate ideas, stimulate analytical approach, encourage critical thinking and lead discussions about all aspects of adapting to new security challenges. This approach includes the provisioning of an adequate scientific database that should provide support to the modern fight against new threats. One of those approaches is this workshop. The need to create a two-way communication between different elements of the scientific community and the management structure of the armed forces with the goal to create an efficient and scientifically based answer to modern security challenges makes us rethink the way that communication between management structures is being made. Political and strategic changes that happened after the Cold War have dramatically changed the global security environment. Traditional challenges that implied competition between super forces have been reduced or have disappeared entirely. Despite the expectations, ending of the Cold War hasn't made the world a safer place to live or made it easier to understand. Considering the fact that countries continuously tend to accomplish their global political or economic goals with the use of force, a whole new spectrum has appeared of nontraditional asymmetrical challenges for security and stability.

It is still unknown how countries today see their security interests in the new global reality, but it is clear that it is necessary to reevaluate questions and factors that can compromise stability and security.

Modern asymmetrical challenges haven't been the core of security sciences nor have they been the focus of scientific research during and at the end of the last century, whilst today the environment, economy, transnational organised crime, terrorism and social questions are being researched in different ways from the security aspect. Reconciliation between the orthodox and modern approach of sciences that have security as their core interest should spark the creation of a new theoretical view on globalisation but also the applicability to respond against modern security challenges, risks and threats.

If you look at literature just 20 years ago, you couldn't find any threats to security but military; today in the security focus are more societal issues such as climate change, environmental issues, emergencies, etc.

It seems that emergencies, here in the Balkans and in the world, have become more frequent and versatile with more serious consequences for people, material goods and environment. An accident happens every day in the world that may turn into an emergency. Emergencies do not care about people, religions or races and do not recognise administrative, state, human or any other borders and limitations. They spread from one state to another, from one region to another, over continents and oceans and to the space even. They may be caused by natural events, or even more often by human activities, on purpose and by accident.

Regardless of all listed causes, emergencies are usually fatal, particularly if a community is disorganised and unprepared for them. It is therefore necessary to create and strengthen community's awareness of possible threats as steps that should be undertaken so that community and individuals are protected; appropriate intervention plans should be developed so that emergency, if not stopped, is at least mitigated and its consequences are reduced as much as possible in terms of lost lives or impaired health and social welfare, damaged property and environmental degradation.

Due to enormous urbanisation, the level of threats to residents and material goods is much higher in urban areas and industrial centres. This is logical since more than 50% of the population lives in towns. Accordingly, the need for a well-organised protection and rescue is the greatest where people are concentrated and significant financial resources have been invested in construction of civil-engineering and industrial facilities. It is hence understandable that huge efforts and great financial resources have been invested today in organisation and preparation of community for emergency management.

Emergencies are characterised by three key factors:

1. Mass, endangering human lives, material goods and environment.
2. Inability or unpreparedness of regular (urgent) services to reach and stop the danger.
3. Consequences of the threat exceed inflicted community's capacities (resources) and require that it undertakes special measures and actions.

From the beginning of the world till today, mankind has been threatened by the phenomena caused by the natural forces which could not be predicted and controlled by humans nor could they influence their appearance, form and intensity of their consequences.

Natural phenomena, such as earthquakes, landslips, floods, storms, epidemics, epizootics and droughts, caused throughout history suffering of the population and huge material destructions. Regrettably, such phenomena are common even today although the development of civilisation enabled the people to live in an easier way.

Threats caused by natural forces are unpredictable, and mankind cannot have any influence on them, on when they will occur, on what form they will take, on what intensity they will possess and even on what their consequences will be.

According to the way they are demonstrated, the natural threats are in most of the cases unexpected threats such as floods, earthquakes and landslips. The floods are the most often form of natural disasters which involve sudden migration of the population and lack of goods, whilst the earthquakes cause large number of the dead and huge damage to the infrastructure.

Many human activities performed by the individual and society, consciously or unconsciously, violate or derange directly or indirectly normal natural processes which endanger human lives, plants and animals. Environmental threats are the processes that occur due to the people's way of living, i.e. their behaviour and activities which harm the nature and its processes. Ecology threats which endanger natural relations refer to natural processes such as change of climate, reduction of the ozone layer in the stratosphere, acidification, loss of biodiversity, pollution of drinking water, pollution of the seas and coasts, deterioration of the forests, degradation of the soil, problems relative to waste management, ecological incidents, chemical risks, threat from the genetically modified organisms, threat from radiation, urban stress, etc.

One can say that emergencies are the result of the performance of numerous natural, human, economic, social, cultural, institutional, political and even psychological factors which determine human lives and shape the environment in which they live. Under such circumstances, the importance of regional cooperation and active participation of the regional organisations would be predominantly in education and prevention of possible threats or in eliminating consequences of such emergencies.

Natural sources of security threats refer to natural and geographic area elements: climate, territory and geographic position.

Number of natural or technical and technological disasters has gone up in the world last couple of years. Consequently, the damages they caused have been much greater. This has been caused by a number of reasons, such as increased population on the Earth and progressive urbanisation of territories, which all lead to anthropogenic influence on the planet's ecology and possible global climate changes on it. Today, we are facing with three groups of ecologic challenges, risks and threats to security: the first group includes interstate conflicts caused by laying claim to or usage of renewable and non-renewable natural resources; the second group refers to residents' movements caused by varied ecologic blasts that may lead to conflict of interests and groups of different ethnic and religious orientation; and, finally, the last group includes falls of social structures and interior societal instability that may bring about violence caused by endangering a greater number of residents by using up certain natural resources, energy or raw materials, climate changes, contamination of environment, etc.[1]

Depending on how they are manifested and what consequences they cause, general characteristics of all emergencies may be summed up as follows[2]:

- Emergencies that put an emphasis on crucial disproportion between current needs for protection and rescue and existing capacities.
- Some of them may occur at certain places or territories (floods), and others may happen anywhere (fires), whilst some are of local character (earthquakes).
- Depending on its intensity, every emergency may cause greater or smaller number of casualties, i.e. jeopardise people's health and degrade living and working environment to a major or minor extent. Almost all emergencies, at relatively limited area, lead to a great number of victims, usually injured or wounded at that moment, qualitatively and quantitatively different from the regular ones.
- Nature of origin, expansion and effects of some threats is such that mankind is not able to defend. However, some threats can be successfully discovered and then partly or fully avoided, mitigated and removed.

---

[1] Thomas Homer-Dixon, 'Environmental Scarcity and Intergroup Conflict' in: Michael T. Clare, Yogesh Chandrani, *World Security – Challenges for a New Century*, Third Edition, St. Martin's Press, 1998. pp. 342–65.

[2] Jakovljevic, V.: Working materials (*Etiology of Threatening People, Material Goods and Environment*), Faculty of Security, Belgrade.

- Different psychological reactions in various manifestation forms follow emergencies and might have a negative influence on organisation and successfulness of protection and rescue.
- Destruction and damage to material resources inflicts those at ambulance and emergency services and other institutions in charge of fighting emergencies.
- Emergencies require urgent interventions, rescue and assistance; hence, automatism and professionalism are very important in doing these.
- Threat is polymorphic. Two same threats, of the same origin and intensity, usually bring about two different situations in terms of overall consequences they cause. This fact makes them individual, which derives from concrete prerequisites related to place and time: structure of settlement, general culture and customs, social services development, wider community structures and its geographic characteristics, town planning solutions, part of day and year, etc.
- They occur at certain points or zones; hence, we face a phenomenon called parallelism. The inflicted zone is a hotspot that is all of a sudden and at no time altered, whereas other parts of the territory remain uninflected and run usual living and working processes.
- A number of emergencies and their characteristics and consequences require the state and society to undertake countermeasures to prevent emergencies and eliminate their consequences and therefore set up its own managerial system – emergency management system.
- We may say that the emergency management system is a very complex process of planning and managing processes, measure and functions, focused on protection of people, material goods and society as a whole, from negative consequences of emergencies. The emergency management system should fulfil criteria that meet requirements for implementation of a dynamic early warning system and provision of appropriate answers for a specific situation at requested dynamics.
- Main action directions have been therefore defined, essential for joint activities: precaution and preparedness, preparation and planning, mitigation, response and recovery.

How to behave when some threats occur is a question looking for an answer for centuries. Such a question is nowadays even more topical, since natural disasters have jeopardised a wide space in the course of last decades. Prevention, avoidance or reduction of possible consequences of unexpected effects imposes efficient intervention issue at all catastrophe-related stages, with the following mandatory requests:

- How can one disclose coming threats and their possible consequences as soon as possible?
- Which bodies and organisations have to react instantly and how can they direct their activities?
- What could be done so that all social entities are ready to react to all threatening issues?

- What is the sequence of joining protection and rescue actions for heterogenic participants?
- Which documents, as operational laws, should be prepared before a catastrophe takes place?
- Which measures and actions are to be undertaken once a catastrophe occurs?

These, and all other requests, undoubtedly stress out the responsibility of protection and rescue participants to get to know sources, causes and consequences of certain threats and find appropriate solutions of organisational, technical, medical and other characters that would efficiently protect inhabitants and their material goods and prevent environmental degradation.

Escalation of issues related to protection of civilian population, material goods and environment in emergencies has made it evident that many, lest we say all related issues, cannot be resolved by isolated actions of individual countries. This is particularly the case when it comes to across-border and global issues.

States are forced to cooperate in harmonisation of emergency management system standards in order to facilitate and make protection and rescue activities more efficient, as well as those protecting material goods and cultural heritage and environment in emergencies. States also need to coordinate adoption of supranational policy measures (e.g. subregional, regional and global); development of international regulations and strategies, programmes and plans aiming at harmonisation of creation; and reinforcement of protection and rescue system in emergencies.

Conclusion

The list of natural threats has grown in the course of time. Civilisation progress and boom have even extended this list. Disasters are more comprehensive and rougher in qualitative sense, and one needs to decipher them first of all. Some threats have been neglected for years, even centuries, depending on education, culture and one's own ability to spot them. It is, therefore, necessary to list them all and classify what many experts and scientists have been doing at the moment.

It is essential we undertake a number of measures, means and procedures in order to have, as much as possible, safe and bright future and reduce potential threats to the lowest possible level. In addition, it is necessary we become accountable for undertaken actions relevant for security of the mankind, which will not be able to do in the nearer future.

A state is obliged to organise people's lives; fulfil their normal needs; guarantee rights and freedoms to everyone, as well as security of its citizens; and take care of equal opportunity provision for all future generations. It has to, besides meeting legal and social aspects, focus on social, natural, ecological and sustainable growth dimensions. Security issues should be the prioritised ones for a country and also how to preserve and improve natural environment and reasonably use natural resources.

Contents

About the Authors

Jeffrey A. Andrews has over 20 years combined professional experience as a university senior research and federal agency scientist. Currently, Mr. Andrews manages the US Africa Command (USAFRICOM) Environmental Security programme. The programme focuses on three elements: preventing or repairing military damage to the environment, preventing or responding to environmentally caused conflicts and protecting natural resources throughout Africa. He also supports the Secretary of Defense and USAFRICOM theatre security cooperation strategies by conducting partner capacity-building engagements. As a senior research scientist of Colorado State University, he oversaw the natural resource programme on over 800,000 ha in Alaska, utilising innovative management strategies in conjunction with other federal and state agencies and tribal entities. As a US Marine Corps encroachment manager and later as an Army Regional Environmental and Energy programme manager, he worked with federal, state, tribal and local groups to ensure proper coordination and consistent application of environmental and energy policies. He has managed diverse programmes with facets focusing on scientific studies, project management, multinational working groups and host nation agencies. Mr. Andrews frequently works with other governments and federal, state and community organisations to build cooperative relationships. He finds communicating across cultures challenging and rewarding and has worked and travelled in over 68 different countries.

Carl Bruch is a senior attorney and codirector of international programmes at the Environmental Law Institute (ELI). He has helped countries and organisations throughout Africa, the Americas, Asia and Europe develop and strengthen their environmental laws, improve institutions and build capacity. His work focuses on environmental peacebuilding (especially after conflict), environmental governance, adaptation and environmental emergencies. He has edited more than ten books and authored dozens of articles. He co-chairs World Commission on Environmental Law's Specialist Group on Armed Conflict and the Environment and is the secretary-general of the International Water Resources Association.

Dr. Imes Chiu serves as an associate professor at the Daniel K. Inouye Asia-Pacific Center for Security Studies. At APCSS, she teaches and conducts research on regional security topics with particular interest on crisis management at the operational and strategic policy level. Prior to working at APCSS, she established the Applied Research Division at the Center for Excellence in Disaster Management and Humanitarian Assistance focusing on civil-military operations where she published several articles focusing on best practices and lessons learned in disaster response. Dr. Chiu has 20 years of professional and academic experience related to stability and support operations in the Indo-Asia-Pacific region. Prior to working at the US Department of Defense, Dr. Chiu taught at Cornell University. She speaks five languages.

Dr. Andmorgan R. Fisher is an environmental scientist at the Geospatial Research Laboratory, Engineer Research and Development Center (ERDC), US Army Corps of Engineers. In addition to her duties at Corps of Engineers, Dr. Fisher currently serves as an affiliate professor at George Mason University in Fairfax, Virginia, and at the Uniformed Services University of the Health Sciences in Bethesda, Maryland. She is also an associate at the Microbiome Analysis Center. Dr. Fisher's primary expertise is in environmental science, bioinformatics, spectral analysis and distributed sensing platforms. She has led multiple basic and applied interdisciplinary research projects seeking improved understanding of the operational environment through the nexus of remote sensing technologies, terrestrial sciences and data informatics. Prior to joining the ERDC, Dr. Fisher served as a professor at both Northern Virginia Community College and George Mason University where she taught a number of courses in the biological and environmental sciences. She formerly has worked as a contractor for the Department of Energy and Department of Homeland Security where she conducted threat analysis and has also served in the US Army as an intelligence analyst.

Vasko Galabov M.Sc., is an associate researcher at the National Institute of Meteorology and Hydrology – Bulgarian Academy of Sciences. He is working in the area of operational marine meteorology. His area of research is marine numerical modelling, studies of storms climatology and marine renewable energies.

Craig Hanrahan is a field service administrator for the Tennessee Emergency Management Agency (TEMA) in Nashville. In this position, Mr. Hanrahan leads 34 field personnel, stationed statewide, responsible for coordinating federal, state and local resources for emergency response and disaster recovery efforts. Mr. Hanrahan has 13 years of total experience at the TEMA, beginning his emergency management career at the agency as an area coordinator responsible for responding to and assisting counties in disasters and emergencies. Mr. Hanrahan's accomplishments at the TEMA have included leading the agency through reaccreditation in 2013 in meeting 64 national standards of emergency management excellence, supporting the TEMA and the state of Tennessee through 19 federally declared disasters and rebuilding the operations and staff of TEMA's 24-h Watch Point into an agency best practice.

Prior to the TEMA, Mr. Hanrahan served 27 months in the Peace Corps as an environment sector volunteer stationed in North Africa where he worked on local projects to improve access to drinking water and life-sustaining resources and assisted in an array of collaborative efforts. Mr. Hanrahan is a member of the American Society for Public Administrators and the Emergency Management Association of Tennessee. He holds a master of science in public administration from Tennessee State University in Nashville and has graduate certificates in health administration and planning and geospatial information systems.

Valentin Kazandjiev is a professor, doctor and, since 2003, head of the Division of Agrometeorology. Prof. Kazandjiev passes specialisations in Meteo-France, Toulouse, France, in 1991 and 1995; International Training Centre, Bet Dagan, Israel, in 1992; International Centre for Theoretical Physics, Trieste, Italy, in 1994; Statistical Services Centre, University of Reading, UK, in 1997; JRC, Ispra, Italy, MARS project, in 1998 and 2003; University of Natural Resources and Life Sciences (BOKU), Vienna, in 2008; and Institut National de la Recherche Agronomique (INRA), Avignon, France, in 2009. Scientific areas, activities and experience include agrometeorology and agroclimatology, agrometeorological forecast and farmer services, investigation of climate change impact on the agroclimatic resources, application of remote sensing observations for agricultural forecasts, modelling and middle- and long-range forecasts for expected yields and managing of automatic meteorological stations ADCON® and DAVIS.

Dr. W. Chris King currently serves as the dean of the US Army Command and General Staff College and chief academic officer of the Combined Arms Center's leader development and education programme. In this position, he oversees a faculty of over 400 in providing professional military education to more than 7,000 students per year from all US services and more than 120 countries. Dr. King was commissioned into the Ordnance Corps in 1972 after completing his B.Sc. in chemical engineering at Tennessee Technological University. In 1974, he completed his M.Sc. in civil engineering (environmental) and entered active duty as a sanitary engineer in the Medical Service Corps. As a junior officer, he completed numerous assignments within the Army's preventive medicine programme and the US Army Corps of Engineers, highlighted by serving as project manager and assistant division commander of the US Army Corps of Engineers Division, Huntsville, AL, where he worked on the Army's chemical weapons disposal programme. He earned his Ph.D. in environmental engineering at the University of Tennessee in 1988. His next assignment was as chief of the Environmental Health Engineering Division, US Army Environmental Hygiene Activity-West, Aurora, CO. In 1991, he was deployed as the officer in charge of the Southwest Asia Health Risk Assessment Team, which determined health risk to US troops exposed to the smoke from the Kuwait oil fires and supported the restoration of Kuwait. In 2000, he completed his M.A. in National Security and Strategic Studies at the Naval War College. Dr. King is a licenced professional engineer and is board certified by the American Academy of Environmental Engineers and Scientists with a specialty in hazardous waste management. He is a founding member of the Global Military Advisory Council on Climate Change.

Anna Kortcheva Ph.D., is an associate professor of marine meteorology at the National Institute of Meteorology and Hydrology – Bulgarian Academy of Sciences. She is head of the Marine Forecast Division. Her scientific and applied research is related to the development and implementation of numerical wave models for operational prediction of wind waves in the Black Sea. She is a member of the Expert Team on Waves and Coastal Hazards Forecast of the Joint Technical Commission for Oceanography and Marine Meteorology of World Meteorological Organization and UNESCO's Intergovernmental Oceanographic Commission.

Lyubka Pashova Ph.D., is an associate professor of geodesy at the National Institute of Geophysics, Geodesy and Geography – Bulgarian Academy of Sciences. Her research interests and activities focus on topics including statistical methods for geodetic data processing, coordinate reference systems and processing and analysis of geospatial information. She is a member of the Bulgarian Cartographic Association, of the Association for Geospatial Information in South-East Europe and of editor's group of two journals: *Journal of Geodesy* and *Micro, Macro & Mezzo Geo Information*. She leads and participated in several projects in the field of geodesy and geodynamics at national and international level. Since October 2014, she is GEO principal alternate of Bulgaria in the Group on Earth Observations.

Georgi Petrov is a graduate fire safety and rescue engineer of the Academy of the Ministry of Interior of the Republic of Bulgaria. Since his graduation, he has been working for 4 years as an expert in the international cooperation, NATO, EU and Humanitarian Operations Department of Directorate General Fire Safety and Civil Protection to the Ministry of Interior of the Republic of Bulgaria. Georgi has gained substantial knowledge on disaster management system and policies in Bulgaria. He possesses an operational experience as in May 2014 he has been deployed for 2 weeks as a deputy team leader of the Bulgarian High Capacity Pumping (HCP) Module which provided assistance to the Republic of Serbia due to the heavy rainfalls which caused excessive flooding in the country.

Stoyan Stoyanov has worked as a senior expert at CMDR COE since 2013. He graduated from the National Military University in 2000 and was working as a logistic officer until 2012. He graduated from the Georgi Rakovski Military Academy in 2013. He holds master of management from the University of Economics – Varna.

Katarina Strbac has a rank of a colonel from the Serbian Armed Forces. She graduated from the National Defense College (1986), obtained her master's degree at the same faculty (2000) and Ph.D. at Belgrade University (2006) and is entitled a Ph.D. of security, defence and protection. Her Ph.D. thesis is dedicated to protection of civilians in emergencies. She has additional education such as Leaders for the Twenty-First Century in Marshall Center (2004), managing defence in democracies in Cranfield University and Defence Academy (2006), CSDP EU seminar (2012) and certain number of courses and seminars concerning security, SSR, EU, NATO and emergencies. From 2005 to 2008, Col. Ph.D. Strbac was head of the Section for Strategic Documents in the Department for Strategic Planning, MoD, Serbia, and from 2008 to 2015 was head of the division for strategic analyses and security integrations, Institute for Strategic Research, Sector for Defence Policy, MoD, Serbia. From 2015 to the present, she is director of the Directorate for European Integration and Project Management, Sector for Defence Policy, MoD, Serbia. She is author of three books from area of security and many articles in Serbia and abroad concerning emergencies, integrations and asymmetric threats to security. She was running several projects such as security and defence aspects of Serbian accession to the EU and security trends in SEE till 2020. She is member of the negotiation team for Chap. 31 of Serbian accession to the EU.

Dr. Marc van den Homberg as an applied researcher and consultant, focuses on how (big) data and information management can improve preparedness and response to natural disasters and complex emergencies. He had short- and medium-term field experiences in Africa (Burundi, Burkina Faso, South Africa) and Asia (Bangladesh, India, Nepal, the Philippines, Vietnam). He holds the disaster management certificate from the International Federation of Red Cross and Red Crescent Societies and an M.B.A. and a Ph.D. in physics. Marc founded within TNO the ICT for Development (ICT4D) team, with which he implemented pro-poor ICT innovations in developing countries through multi-stakeholder projects. Marc lived and worked in Belgium, Germany and the USA and currently lives in France.

Janani Vivekananda is the head of environment, climate change and security at International Alert where she is responsible for research and implementation support on climate change, environmental and natural resource-related dimensions of peacebuilding and security. Her role involves leading on innovation, analysis and documentation of new research and liaising with academic, policy and practitioner communities on climate change, development, peace and security. She has been working on climate change and security issues since 2006 and has published widely on the subject. Her specific interests include the implications of climate change policies on peace, the links between climate change and community resilience and opportunities for peace-positive responses to climate and environmental change and disasters. Janani's previous roles include disaster risk reduction advisor at Plan International UK and security and peacebuilding programme officer at International Alert and consultancies on conflict analysis and aid effectiveness for SNV Nepal and UNDP Nepal. She has an M.Sc. in violence, conflict and development from SOAS and a B.A. in politics, philosophy and economics from New College, Oxford.

Agenda

Orlin Nikolov and Swathi Veeravalli

Advanced Research Workshop Proceeding 2016
Implications of Climate Change and Disasters on Military Activities: Building Resilience and Mitigating Risks and Vulnerability in the Balkans
5–7 July 2016
Sofia, Bulgaria

Valentin Kazandjiev – *Climate Change – Fundamentals, Agroclimatic Conditions in Bulgaria, and Resilience Agriculture Through Adaptation*
Swathi Veeravalli – *Perceptions on the Current State of Human-Environmental Security in the Balkan Region*
Chris King – *Civil-Military Operations, Climate Change, and Disasters*
Stoyan Stoyanov – *Climate Change Implications on Military Activities*
Janani Vivekananda – *How Are Climate Change and Human Security Interrelated*
Jeffrey Andrews – *Perspectives from Africa Command on Linking Human and Environmental Security*
Carl Bruch – *Disasters and Security*
Georgi Petrov – *Bulgaria 2015 Peer Review Report*
Janani Vivekananda – *DRM: Using a Resilience Systems Approach to Plan for Multiple Stressors*
Doug Brantley – *A Military Perspective for Supporting a Whole of Government Approach During Disaster Planning and Response*
Luan Qafmolla – *Spent Hight Activity Radioactive Sources*

O. Nikolov (✉)
CMDR COE, Sofia, Bulgaria
e-mail: orlin.nikolov@cmdrcoe.org

S. Veeravalli
Geospatial Research Laboratory, US Army Corps of Engineers, Washington, DC, USA

O. Nikolov, S. Veeravalli (eds.), *Implications of Climate Change and Disasters on Military Activities*, NATO Science for Peace and Security Series C: Environmental Security, DOI 10.1007/978-94-024-1071-6_1

Michael Campbell – *Examples of Open Source Software and Imagery for Regional Land Cover Mapping*
Imes Chiu and Kosta Delev – *Facilitated Discussion: Lessons Learned and Best Practices in Complex Catastrophes from the Indo-Asia-Pacific Region, Perspectives from the Affected and Assisting States*
Carl Bruch – *Assessing and Leveraging Existing Capacity*
Marc Van Den Homberg – *Identification of Data Requirements*
Katarina Strbac – *Serbia, Bosnia, and Croatia 2014 Floods*
Andmorgan Fisher – *Remote Sensing to Assess Effective HADR Operations*
Craig Hanrahan – *The Condition and Impacts of Climate Change*
Swathi Veeravalli, Stoyan Stoyanov, and Katarina Strbac – *Facilitated Dialogue: Sustaining the Community of Interest Within the Balkans*

Crisis and Disaster Management Terminology

Stoyan Stoyanov

Abstract Terminology refers to terms and their use. In the other hand, terms are words, composite words and/or multiword expressions that in specific contexts are given specific meanings. The context defining is vital to be done in the very beginning of the terms usage, while the article terms usage context will be specified later on. Every single subject area terminology is a toolbox that the subject matter experts (SMEs) in that field use in order to communicate and to understand each other. It sounds logical but the real life is not like this, and C&DM field is not an exception – using of all main terms in it is not unique, unambiguous, and clear.

Keywords Crisis and disaster management • National security protection system • Hazard • Threat risk

Introduction

Do you know what main terms in crisis and disaster management (C&DM) field exactly mean? If "Yes," do you believe that your counterparts have the same understanding to them? If "No," do you think that you speak one and the same language? I, personally, would answer "No" to all of these questions and because of that the present article was written.

Terminology refers to terms and their use. In the other hand, terms are words, composite words, and/or multiword expressions that in specific contexts are given specific meanings. The context defining is vital to be done in the very beginning of the terms usage, while the article terms usage context will be specified later on. Every single subject area terminology is a toolbox that the subject matter experts (SMEs) in that field use in order to communicate and to understand each other. It sounds logical but the real life is not like this, and C&DM field is not an exception – using of all main terms in it is not unique, unambiguous, and clear. One of the main reasons is the complexity of the matter and lack of consensus into the academia and among the different institutions dealing in the field for the meaning of main terms

S. Stoyanov (✉)
CMDR COE, Sofia, Bulgaria
e-mail: stoyan.stoyanov@cmdrcoe.org; stoyanov.web@gmail.com

O. Nikolov, S. Veeravalli (eds.), *Implications of Climate Change and Disasters on Military Activities*, NATO Science for Peace and Security Series C: Environmental Security, DOI 10.1007/978-94-024-1071-6_2

and definitions. We should also take into account the fact that all the abovementioned institutions are relatively conservative and inert in regard to any changes. Quite often, they follow complex interests and, respectively, consensus achievement is hard even impossible. Good example is North Atlantic Treaty Organization (NATO). In its Glossary of Terms and Definitions (AAP-06, 2015), the alliance has not pointed out a definition of one of the main terms the organization uses related to one of its core tasks, namely, *crisis*. This document gives a definition for *crisis management* and is quite logical to ask "how to manage something that is not clearly defined." I personally, taking part in various discussions, mainly into security field, ask for the definitions that my collocutor(s) uses in his/her work, and till now I have not received a clear answer. It could be understood pretty well because there is no a single document that describes all main C&DM definitions in clear and logically connected manner.

There are fractional exceptions to this, and a good example is the UNISDR.[1] Terminology. It aims to promote common understanding and common usage of disaster risk reduction concepts and to assist the disaster risk reduction efforts of authorities, practitioners, and the public.[2] The document gives definitions for main terms, but there are some discrepancies and overlapping which are confusing to me. The new version of the document, or the group that develops the new edition, promises that it will be so precise as to meet main requirements for at least the next decade. However, reading draft version of the document, which will be adopted by the end of 2016, the same mixed approach could be seen. The main weakness could be found there is that no one tries to point out a specific context (point of view/ perspective) and to try to connect terms with each other. In this regard, the next few pages could be seen as a possible solution and does not pretend to be exhausted in its content but to set a solid basis for thinking, discussion, and debate.

A New Story of Tower of Babel

Before we start with term explanations, we should set out a perspective or point of view – in other words, in which **context** C&DM terms will be discussed later on. It is substantial because it is a standpoint – our "ground zero" in this article. Discussing all of the terms and their definitions, we should assume that we do that from governmental, national, and managerial level. In other words, how different events influence the governance and continuity of government and its mechanisms of power applying; how structures of power interact with the security environment – how such events affect national security level.

Security/national security is the first term I am going to discuss and deem fundamental in the field due to the fact it gives a basis to the topic. Security is a dynamic state in which system (total of interacting parts) existence and interests are

[1] The United Nations Office for Disaster Risk Reduction.

[2] UNISDR Terminology for Disaster Risk Reduction, 2009.

guaranteed. This system must be alive with all its characteristics – to exist following its own regulations, to be capable to provide a feedback, and to be able to adapt itself to the environmental changes. In this case, the system should be considered as a group of interacting people – a society. In order to be able to maintain the national security at desired state, the government must develop and maintain a *national security protection system*. This protection system should encompass at least (1) managerial bodies, (2) executive bodies, (3) coordination center(s), and (4) communication and information system (CIS).

Normality/new normality is a state where the level of security is at an acceptable level, and it (the security level) increases to some extent.

A **hazard** should be considered as an event that has the potential to disturb the normality in general. This term should be used to point out all different occurrences that have a potential to cause disturbance into normality. Due to the fact that hazard is closely related to threat, they are both connectively described into next paragraph.

Threat and hazard should not be considered as one and the same, but often they are incorrectly used interchangeably. Hazard could be deemed as a source of a risk. A hazard is a potential threat. The hazard is a potential in a not hurtful state; the threat is the same potential in its damaging state. For instance, a dam is a hazard as long as it is in its optimum capacity, but it becomes a threat when it is full of and threatens to overflow and flood a vast region.

In the other hand, **risk** is a characteristic of the threat. It gives qualitative and quantitative features of the threat. The risk, based on analysis, connects the threat impact/consequence and likelihood/probability. It could be illustrated by following equation:

$$\textbf{Risk} = \textbf{Impact} \times \textbf{Likelihood}$$

Before discussing the next three terms – resilience, vulnerability, and adaptation – it is worth to mention their similarity to hazard, threat, and risk. They are closely related and have much in common.

In general, **resilience** is an ability of a system to cope with a change – to rebounce to a new normality. Moreover, in the context that we use, an emphasis should be put on the fact that (1) the system is alive and (2) the change has more or less threat characteristics – this fluctuation of normality has negative features.

Vulnerability refers to an inability of a system described so far to withstand the effects of an environment which would decrease its security level.

Adaptation, by itself, is a connection between resilience and vulnerability, and to be more precise, it is an inversely proportional construction. As much the system resilience increases that much vulnerability to events negatively influencing security level decreases. The adaptation is related to system's behavior – its policy to a problem. Due to the fact that the system security has extremely high priority, the adaptation to the environment should possess high priority as well (Fig. 1).

The next two terms, **capacity** and **capability**, are also closely related and should be considered in their interrelation, but both should be distinguished from each

Fig. 1 Connection between vulnerability (V), resilience (R) and adaptation (A) (Source: the author)

other. First of all, building capacity and capability is an activity that should be done in preparedness phase of disaster management process (this process is to be discussed later on) based on all findings made during the risk assessment/prevention/mitigation phase of the abovementioned process.

The **capacity development** should be considered more in general and should be an attribute of the society and its ability to be resilient to negative changes in the security environment – a societal coping ability. Along with many other attributes such as infrastructure and physical means, it must include human knowledge, skills, and attitudes, as well as collective attributes such as social relationships. In the other hand, the **capability development** should be an activity related to the government instruments to cope with the security stressors. Capability development should be considered as a process which the government conducts during recovery, mitigation, and preparedness/preparation phases of disaster management process. Usually, it consists six phases – strategic environment assessment, identifying capability needs, capability requirements deriving (capabilities needs qualitative and quantitative characteristics), conducting of a gap analysis, finding possible solutions (using DOTMLPF-I[3] Approach), and applying of the new capability.

Incident, no matter industrial, transport, informational, technological, political, etc., should be considered as an event which is confined in its duration which has potential to cause an emergency, disaster, crisis, or even catastrophe. The incident could be human instigated or due to technical failure – caused intentionally or unintentionally.

The next three terms emergency, disaster, and crisis which I am going to focus on are fundamental in CM&D field and have to be clearly defined and used in specific conditions – not to be overlapped or incorrectly used in one and the same meaning. They are also closely related and should be taking into account their interrelation. These three different dynamic conditions (emergency, disaster, and crisis) could be caused by various reasons. They could be summarized in four distingue groups – natural, technological, human instigated/induced, and complex (combination of the previous three mentioned). These four groups of factors can be divided in two subgroups – sudden-onset (short-term development from hours to weeks) and slow-onset states (long-term development from years to decades). No matter of their time characteristics, the basic difference between emergency, disaster, and crisis should be sought in the level of using of capacity/capabilities developed in disaster

[3] DOTMLPF-I stands for Doctrine, Organization, Training, Material, Leadership (including Education), Personnel, Facilities and Interoperability.

management preparedness phase, their sufficiency to cover the basic needs, and how the event or/and the process influences the security level.

Emergency is a dynamic condition in which the normal state is disturbed, and the level of the system's security decreases. First, this condition is dynamic due to the fact that many factors have influence to it, and it could go to a new normal condition, could stay stable, or could deteriorate and become a disaster. Second, the government answers to the emergency using the capabilities developed during the preparedness phase and supported by the societal coping capacity. Third, the means to answer to the irregularity that society possesses are *enough* to cope with it. The society uses preliminary spared resources to cope with that unacceptable condition, and the most significant in that case is that it could go back to normality using own means only. Furthermore, **disaster**[4] is a dynamic condition in which there is a human suffering, human lives are threatened or/and lost, significant economic (material and/or environmental) losses are caused, and the community is *not able* to cope with that negative dynamic condition in order to decrease human suffering as well lives and economic losses. Disaster, similar to emergency, could be at local/municipal, regional, or national level, and managerial body could be a concrete mayor, a regional government/council, or a national government. This article is set on national (governmental) context as the government represents the highest and last resort of power of a country as an independent organization. Because of that I would not use small-scale or large-scale disaster – the only criterion is governmental ability to cope with it using owned capabilities and societal capacity. In case of disaster, additional resources not possessed by the government are essential, and the government that has to deal with such undesired state condition should use various foreign tools for support already established. **Crisis**, by itself, is a dynamic condition in which the national security level is at critical level, and continuity of the government is threatened. In this case, negative feedback level is extremely high, and the society searches improvement through change in its governance.

Humanitarian crisis addresses a situation in which a group of people connected on their specific characteristics are subjected to conditions in which their fundamental human rights are violated. In that case, there are failed governmental, no matter homeland or foreign, efforts to alleviate the situation. Respectively, in case of a humanitarian crisis, a humanitarian assistance is required. It is always provided by international community due to the fact that the respective government is not able/willing to ensure the resources required – often, its policy intentionally causes the undesired state.

Catastrophe should be considered as a condition which affects vast area and more than one country, in which the national, regional, and/or global security is at extreme low level, and the system's existence is significantly threatened. It could be

[4] A serious disruption of the functioning of a community or a society involving widespread human, material, and economic or environmental losses and impacts, which exceeds the ability of the affected community or society to cope using its own resources – UNISDR Terminology on Disaster Risk Reduction definition, 2009.

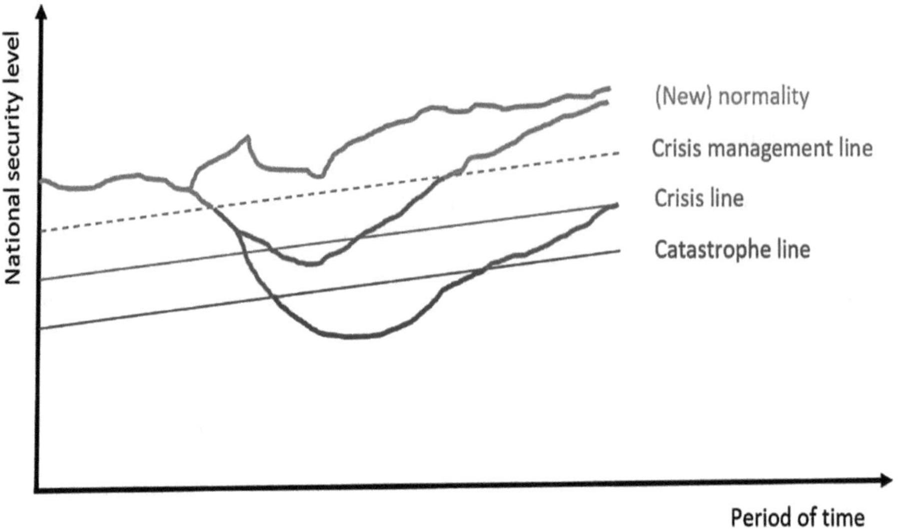

Fig. 2 Crisis and catastrophe connection (Source: the author)

caused by natural hazard (massive volcano eruption), technological incident (Chernobyl NPP meltdown), and human instigated (civil war in Syria) (Fig. 2).

Complex emergency (widely used term) and **protracted crisis** (relatively new term) stand for one and the same dynamic condition caused by various and intricately linked (natural, technological, and/or human instigated/induced) factors in which the respective government is not able to cope with the situation, and its continuity is highly threatened. In that regard, both terms are relatively incorrect and controversial. More precise term to be used in such a situation is **complex crisis**.

Emergency/disaster and crisis management processes are governance approaches to direct the unacceptable conditions that may appear in certain desired direction in order to achieve new stable/normal security environment/level. Emergency (disaster and crisis) management process, through risk management process,[5] should be considered as an organization of responsibilities for resources management in addressing, before, during, and after, an undesired condition of the security environment that affects/decreases national security level.

Emergency/disaster management process encompasses four (illustrative) phases – prevention/mitigation, preparedness, response and recovery. Based on the risk assessment which includes risk identification, analysis, and evaluation, concrete actions are taken for some of the hazards to be prevented. For those that cannot be prevented, the risk level should be decreased to tolerable level.

[5] For more information, see Risk management — Principles and guidelines, International Standard ISO 31000, 2009.

Fig. 3 Emergency/disaster management process/cycle (Source: UN Global Assessment Report on Disaster Risk Reduction, 2015)

For all hazards with risk level above the acceptable one, the government should develop capability and respective coping capacity (see capability/capacity development) – achieving a state/condition of disaster response preparedness.

The response phase could be divided into following subphases – saving people, providing immediate assistance, assessing main damages, and ongoing assistance and restoration of essential infrastructural services. It is worth to mention that response phase starts before disaster to strike – early warning could be crucial for a disaster or even an emergency prevention. Recovery phase could encompass reconstruction (resettlement and relocation) and rehabilitation (economic and social recovery). Vital for the new normality achieved, based on lessons learned and using of new disaster management approaches, is to set state for increasing the national security level (Fig. 3).

Furthermore, crisis management process should be focused on risk assessment (periodically conducted), preparation (based on risk assessment results), prevention/response, and revising/rebuilding phase. As mentioned above, emergency/disaster and crisis management processes, per se, are risk management processes, and they, both processes, overlap in complex manner. Crisis management is closely related to national security, and its main concern is to restore the national security level at new acceptable state. To do that, the respective government should develop proper capability – a *crisis management system* which must be able to deal with different crises in different conditions. The abovementioned system should consist the same element to the system responsible for protection of national security.

Conclusion

In conclusion, I would point that every single nation as a live system should build its own Tower of Babel (reliable national security protection system) in order to be able to protect its own interests in ever-changing security environment. It should be achieved through strong political will, clear vision where the nation should be in predefined period of time, pure concept and strategy of what must be done, and precise programs and plans on how all this to be achieved. In order to attain such a vital project, all decision makers and SMEs should use one and the same language – terms and their meaning. All main terms in C&DM such as normality, security, national security, hazard, threat, risk, vulnerability, resilience, adaptation, capacity, capability, incident, emergency, disaster, crisis, catastrophe, humanitarian crisis (assistance), complex crisis emergency/disaster, and crisis management are closely related and should be taken into consideration and comprehended in their complex interrelation, something that, I believe, this modest entry facilitates.

Toward a Balkans' Data for Disaster Management Collaborative?

Marc van den Homberg

Abstract Climate change adaptation and disaster risk management in the Balkans require strong regional cooperation, given that disasters in the Balkans are often cross-border. However, currently information and collaboration gaps occur. This paper proposes to harness the potential of information management, geoinformatics, and big data to bridge these gaps and to create data preparedness as follows. First of all, map regularly regional, national, and local data sets on multi-institutional information needs. Secondly, use new digital and collaborative tools, including geospatial sharing platforms and OpenStreetMap volunteers for sharing, collecting, and using data. Thirdly, build capacity through, for example, courses, exercises, and regional demonstrations. Overall, it is about creating a Data Collaborative for the Balkans, a lightweight way of institutionalizing regional information sharing and creating trust. The NATO Crisis Management and Disaster Response Centre of Excellence can play a pivotal role by convening stakeholders, leveraging military capabilities, and offering an advanced curriculum.

Keywords Information management • Geoinformatics • Data preparedness • Big data • Disaster risk management

Introduction

Disasters in the Balkans are often cross-border due to the size of the countries and the geography of the region [16]. As a consequence, climate change adaptation (CCA) and disaster risk management (DRM) require strong regional cooperation. However, both the WMO Multi-Hazard Early Warning Systems and Risk Assessment in the Western Balkans and Turkey as performed in 2012 and the more recent—2015—EU peer review of the DRM system of Bulgaria [5] identified several shortcomings in this cooperation. This presentation discusses these from a data perspective. Getting the right information to the right people and organizations at the right time is crucial. It requires complete, timely, accurate, valid, and reliable data throughout

M. van den Homberg (✉)
Data4Resilience, Clermont-Ferrand, Auvergne-Rhône-Alpes, France
e-mail: marcjchr@gmail.com

O. Nikolov, S. Veeravalli (eds.), *Implications of Climate Change and Disasters on Military Activities*, NATO Science for Peace and Security Series C: Environmental Security, DOI 10.1007/978-94-024-1071-6_3

all phases of CCA and DRM. For example, in the mitigation and preparedness phase, data is needed for risk assessments and for monitoring progress on the Sendai framework for Disaster Risk Reduction targets; in the preparedness phase, for determining and issuing early warnings; in the response phase, for damage and needs assessments; and in the recovery phase, for recording disaster losses. As compared to DRM, CCA requires data over a very long time frame to monitor global and regional trends. Ideally, adequate *information management*, including the people, technology, and process side, and its closely related *data management*, is implemented. Data goes through a cycle of acquisition, sharing, processing, analyzing, and disseminating between and across stakeholders. In reality, *collaboration gaps* compromise this cycle and cause hiccups. They occur at different levels: between the different stakeholders and data providers involved (ministries, NGOs, private sector); between the international, national, regional, and local level; within an organization (headquarters versus the field); and in time, for example, between the response and recovery phase. In addition, and largely as a consequence of these collaboration gaps, *information gaps* occur. These gaps are bidirectional: sufficiently granular local information is often not available at all or not in time, and important information does not reach those who need it (often the digitally excluded). This presentation addresses how these information and collaboration gaps can be bridged and puts up for consideration a possible way forward for the Balkans leveraging cutting-edge developments in information management, big data, and geoinformatics and drawing from best practices in the humanitarian community.

Characterizing the Information and Collaboration Gap

In this paragraph we describe a generic approach of how to characterize the information and collaboration gap and demonstrate how it works for one specific case. The gaps play out in all phases of CCA and DRM, but are most pronounced in the preparedness and response phase. In these two phases, there are three main groups of individuals and organizations involved [4]: the community affected by the disaster, the professional, and the responding community. The responding community consists of trained disaster management volunteers or all those others in a community who spontaneously offer help and are active in the response. The professional community consists of all those professionals involved in response activities such as the military, civil defense, fire brigade, medical personnel, and government officials. All these three groups have to take decisions in terms of how to prepare themselves or how to respond. Hence, the decision maker has information needs that change over time, and he/she usually requires more detail and accuracy once the response activities unfold. Data sets and information products are or become available as well during or after the disaster hits. The better the match between these needs and data sources is, the higher the *situational awareness* of the decision makers. This improves their *sensemaking* and allows them to take better decisions. Figure 1 shows this as a rather linear process; it is however in reality a highly iterative and

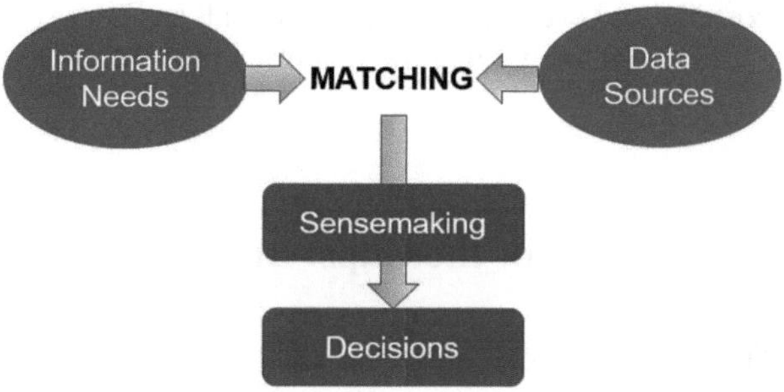

Fig. 1 The schematic relation between matching information needs with data sources

chaotic process where decision makers face compressed timelines and high levels of uncertainty. There are also decision making forces at play that arise from a political and military context rather than only from the humanitarian context. For example, there are cases known where the government demanded actual numbers to be adjusted so that the food security situation was underreported and an intervention could be avoided.

This matching approach was followed in a case study of the 2014 floods in Bangladesh that affected nearly two million people and nine districts [12]. Representatives from the three groups described before were interviewed to distill their information needs and to get a first overview of available data sets. Desk research and literature study were used to complete the overview of the data sets. The information needs could be broken down into two categories, i.e., crisis impact and operational environment, in line with OCHA's Multi-Cluster/Sector Initial Rapid Assessment analytical framework. Crisis impact consists of baseline information about the vulnerability and livelihood of communities in hazard-prone areas, damage and needs assessments, and information about the disaster situation such as which areas are inundated and how long the flood will last. The operational environment comprises coordination and capacity information as well as information on the degree of humanitarian access. In total 71 information needs and 15 data sets with the order of 40–60 indicators each were identified. This specific case study showed that only 27% of the information requirements were met in time (and 62% if timing constraints were not taken into account). Several of the interviewees mentioned that having an overview of relevant data sources and knowing where they could be found could have been very beneficial during the response. More often than not, data critical for the response is collected, stored, and managed outside the responding organizations.

A similar approach can be used for the Balkans. In a case study by Zurich Insurance on the 2014 floods, Zurich [17] concluded regarding information on the disaster situation: "early warnings were not received in time and/or the warnings were not well-enough targeted to the people who should have received and understood them." The EU peer review of DRM in Bulgaria found that "The completion of the risk assessment appears to be hindered by data compatibility issues and bottlenecks in data sharing among the various organizations" [5]. Overall, it is to be expected however that the gap for the Balkans is less pronounced than in Bangladesh, given among others that the countries in the Balkans have a low or below average data poverty.[1]

How to Narrow the Gaps: Building a Data Collaborative

The question now is—once the gap has been characterized—how to narrow it. How to avoid that many organizations are searching independently for the same data, leading to duplications of effort and incomplete and inconsistent databases? How to go from a messy process to one where lots of organizations share information about the data they need and the data they have, resulting in greater access to more consistent data [2]? Slightly different approaches can be used for the mitigation and preparedness versus the response phase. For both phases it is key to create a proper understanding among the different stakeholders of their respective mandates and responsibilities and how this impacts the degree to which they have and can share data.

Once this is mutually understood, activities such as regional demonstrations and exercises can be used to create trust networks before a disaster hits. In parallel, digital and collaborative tools can be used to strengthen existing and create additional networks. OCHA recognized already in 2011 the potential of these tools (at that time still mostly limited to Skype, Google Docs, and Sheets) for so-called Coordinated Data Scrambles (CDS) (see Fig. 2) in the response phase and created a community of interest for CDS [10] to push them forward. Since then the CDS

Fig. 2 Sketch of how a coordinated data scramble (*on the right*) can change the situation of today (*on the left*). *ORG* refers to organization and *DS* to data set [10]

[1] The Data Poverty Index is determined by Internet speeds, computer owners, internet users, mobile phone ownership, network coverage, and higher education. https://www.researchgate.net/publication/283710771_Quantifying_and_Mapping_Global_Data_Poverty/figures?lo=1 gives the scores for the Balkans. There was insufficient data to calculate the Bangladesh score, but it will be the same or worse than for India (which is above average data poverty).

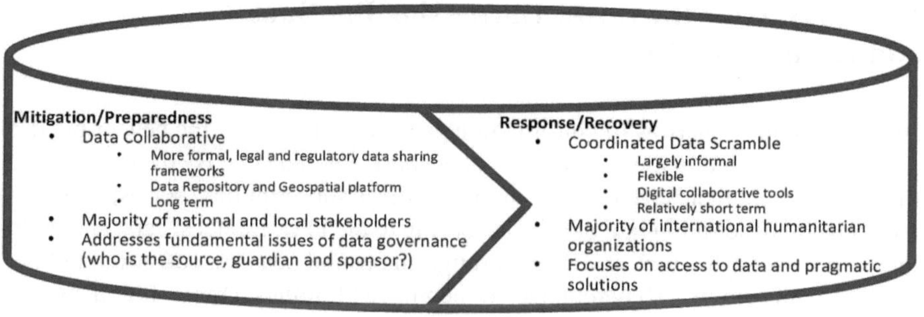

Fig. 3 Data coordination and collaboration continuum

approach has been applied and taken further at the same time in five major humanitarian crises. A coordinator was hired in 2014 to run a CDS for the West Africa Ebola crisis with over 200 participants using multiple coordinating sheets and skype channels for focused conversations. For the European refugee crisis, there were also 200 participants in 2015 using seven channels in one Slack group, and for the recent Ecuador earthquake in 2016, over 700 participants used Telegram. Issues to tackle are whether to make a CDS public or by invitation only and how to avoid too much information, making the CDS a firehose. Solution directions point toward establishing codes of conduct for participants, focusing on (predefined) priorities, and developing common terminologies, such as the Humanitarian Exchange Language ("a simple standard for messy data") for data sets or common categories for information needs (as introduced above).

Whereas the CDS is a largely informal and agile way of collaboration and coordination, a Data Collaborative is a more formal and long-term collaboration between participants from different sectors—including private companies, research institutions, and government agencies—to exchange data to help solve public problems [13]. A Data Collaborative is a very adequate mechanism in the mitigation and preparedness phase to make sure essential baseline data is pre-staged before a disaster hits. Figure 3 shows the data collaboration and coordination continuum, where a Data Collaborative can turn into a CDS once a disaster hits and—the other way around—a CDS can evolve into a Data Collaborative if stakeholders in a CDS realize in the recovery phase the benefits of a more institutionalized way of data sharing. In case of a level 3 disaster, the majority of participants in a CDS consist of the international humanitarian community given that the host nation has asked for international assistance, whereas in a Data Collaborative it consists of national stakeholders.

How to Narrow the Gaps: Harnessing Big Data and Geoinformatics

Another avenue to narrow the information gap is to make use of the new or big data that is being created in the digital age that we have entered [11]. By the end of 2015, there were more than seven billion mobile cellular subscriptions, corresponding to a worldwide penetration rate of 97% according to the ITU. Devices, cars, and buildings can form an Internet of Things through the use of sensors, software, and networked connectivity. Hyperconnected people and things create big data at a fast-paced rate, streaming, or in batches. It ranges from transactional data, captured data (where one can "opt in" or "opt out"), social media, and sensor data up to biological and public records [14]. In fact, digital traces are left behind in almost any transaction, and more and more citizens generate their own online content. These traces are used of course for commercial purposes, but they can also be used for societal benefit. A consortium led by the Flowminder Foundation managed to make use of longitudinal data from six million mobile phone users to unveil hidden migration and mobility patterns in climate-stressed regions of Bangladesh [8]. In contrast to analyzing single-source data, the fusion of data coming from multiple sources, both "new" and "old" data, holds the most promise. Hereby also geoinformatics comes into play, where ICT is coupled with Geographic Information Systems (GIS), GPS, and remote sensing. As an example, Jongman et al. [7] investigated whether combining satellite and Twitter signals could be used for flood early detection in the Philippines and Pakistan. It turned out that the satellite signal was better suited for monitoring large floods and Twitter for floods of any size, provided that affected people in the flooded area had access to social media and were also really using it to share their observations. Similarly, smart social media filtering can be used to collect ground truth information from flood-affected communities as is done with the FloodTags application in Indonesia [6]. Another example of using GIS and GPS is OpenStreetMap (OSM).[2] There is a specific community that works on humanitarian tasks, and this community was activated on request of OCHA during the floods in the Balkans in 2014 [15]. Combining the activities of a remote mapping community with a local one has proven to be the most powerful to resolve specific information requests. Most Balkan countries have an OSM community that is comparable in size to other Western European countries [1], and government agencies can benefit from them, possibly in a similar way as OCHA does. Lastly, geospatial data sharing platforms get more and more adopted and accelerate the use and sharing of open data. The EU enforced the Infrastructure for Spatial Information (INSPIRE) Directive, which led to an EU-wide geoportal [3]. The World Bank runs the Open Data for Resilience Initiative [9] and deploys GeoNode as an open data sharing

[2] OSM is a community of mappers that contributes and maintains data about roads, trails, cafés, railway stations, and much more, all over the world by, for example, going out with a GPS phone or by mapping features on satellite images.

platform in several countries, including a project in Serbia. OCHA created the Humanitarian Data Exchange, where HDX aggregates, stores, and transforms data for the humanitarian community, and the Humanitarian OSM Team created Open Aerial Map providing a simple way to search, share, and use open imagery.

Overall, the use of big data for CCA and DRM is still mostly in the research phase, and (near) real-time implementation is not yet feasible. However, the geospatial sharing platform and OSM mapping applications can and are already directly put to use.

Conclusions

Information management should be a general priority for CCA and DRM in order to close the information and collaboration gap. We propose to focus on data preparedness before a disaster hits:

1. Map regularly regional, national, and local data sets on multi-institutional information needs.
2. Deploy new digital and collaborative tools, including geospatial sharing platforms and OSM to share, collect, and use data.
3. Build capacity through, for example, courses, exercises, and regional demonstrations:
    (a) Create a proper understanding among the different stakeholders of their respective mandates and responsibilities and how this impacts the degree to which they have and can share data.
    (b) Adopt standards and controlled vocabularies.
    (c) Develop information management, geoinformatics, and big data skills.

In the end, it is about creating a Data Collaborative for the Balkans, a lightweight way of institutionalizing regional information sharing and creating trust. It might be advisable to start the collaborative within one country among different institutes, then bilaterally and finally multilaterally around a well-defined data need such as for early warning systems. In this way, success stories can be created that demonstrate the added value of a collaborative. The NATO Crisis Management and Disaster Response Centre of Excellence can play a pivotal role by convening key stakeholders, leveraging military capabilities,[3] and developing and offering an advanced curriculum.

[3] Such as those from the military intelligence and remote sensing community as well as expertise around environmental security.

References

1. Arsanjani J, Zipf A, Mooney P, Helbich M (eds) (2015) OpenStreetMap in GIScience, experiences, research, and applications. Springer International Publishing, Switzerland, pp 1–15
2. Campbell H (2016) The coordinated data scramble for humanitarian emergency response, https://www.youtube.com/watch?v=HmsF743VRX0
3. Cetl V, Tóth K, Abramić A, Smits P (2013) Report on the status of INSPIRE in the Balkan countries. EU Joint Research Centre. http://inspire.ec.europa.eu/documents/INSPIRE_/JRC86293_2013_Report_NSDI_Balkan.pdf
4. COBACORE deliverable D1.1, Scope and Requirements (2014): URL = http://cobacore.eu/vde/cobacore-innovation-station/progress-so-far/public-deliverables/D1.1, Scope and Requirements (15 October 2014)
5. EU (2015) Bulgaria 2015 Peer review report, EU cooperation on civil protection and DRM
6. Floodtags (2014) Using Twitter to get ground truth on floods: an interview with Floodtags founder Jurjen Wagemaker. http://unglobalpulse.org/floodtags-interview
7. Jongman B, Wagemaker J, Romero BR, de Perez EC (2015) Early flood detection for rapid humanitarian response: harnessing near real-time satellite and twitter signals. ISPRS Int J Geo-Inf 4:2246–2266
8. Lu X, Wrathall DJ, Sundsøy PR, Nadiruzzaman M, Wetter E, Iqbal A, Qureshi T, Tatem A, Canright G, Engø-Monsen K, Bengtsson L (2016) Unveiling hidden migration and mobility patterns in climate stressed regions: a longitudinal study of six million anonymous mobile phone users in Bangladesh. Glob Environ Chang 38:1–7
9. OpenDRI (2016) https://opendri.org/project/?mapregion=&mappilar=open
10. Verity A (2016) I scramble, you scramble, we all scramble. http://blog.veritythink.com/post/145957370879/i-scramble-you-scramble-we-all-scramble-for
11. Van den Homberg M (2016) Data for disaster management – mind the gap. Essay in the responding to natural disasters in the MENA region and Asia: rising to the challenge? Series of the Middle East Institute. http://www.mei.edu/content/map/data-disaster-management-mind-gap
12. Van den Homberg M, Monné R, Spruit M (2016) Bridging the information gap: mapping data sets on information needs in the preparedness and response phase. In: Tech4Dev UNESCO online conference proceedings
13. Verhulst S, Sangokoya D (2015) Data collaboratives: exchanging data to improve people's lives. The GovLab. https://medium.com/@sverhulst/data-collaboratives-exchanging-data-to--improve-people-s-lives-d0fcfc1bdd9a#.ctcjbvp2a
14. Whipkey K, Verity A (2015) Guidance for incorporating big data into humanitarian operations. http://digitalhumanitarians.com/sites/default/files/resource-field_media/IncorporatingBigDataintoHumanitarianOps-2015.pdf
15. Wiki OSM (2014) http://wiki.openstreetmap.org/wiki/2014_Southeast_Europe_floods
16. WMO (2012) Strengthening multi-hazard early warning systems and risk assessment in the Western Balkans and Turkey: assessment of capacities, gaps and needs
17. Zurich Insurance (2015) Balkan floods of May 2014: challenges facing flood resilience in a former war zone. In: Flood resilience review

Modern Global Problems Related to Climate Changes and Their Impact on South Caucasus and Georgia

Nika Chitadze

Abstract In the research, the most important problems are analyzed, related to the climate changing and global warming, which are caused by the decreasing area of forests, soil degradation, air pollution, irrational consumption of energy resources, and global warming. Together with the global problems, in the research paper, the impact of the ecological problems on the South Caucasus Region and Georgia is discussed about. The conclusion and recommendation part is dedicated to the several aspects, such as how to avoid ecological problems and which potential has such authoritative international organizations, like NATO.

Keywords Climate • Ecology • Environment • Global warming • Pollution • NATO

Introduction

The most important global problems before the humanity have been especially emerged from the second half of the twentieth century. From one side, in the beginning of the twenty-first century, the civilization's development accessed the high level, but, from the opposite side, the scientific-technical revolution, postindustrial development, and rapid growth of the world population have required the increasing of the exploitation and production of the natural resources.

The global problems (ecologic, disarmament, peace and security, demographic, energy, food, poverty, etc.) concern the whole humanity, and their resolution needs the efforts of the all countries and nations. Each global problem is interconnected and creates the entire chain, where it determined the interrelations between environment and population's social and economic development.

One of the most important global problems is represented by the relations of the humanity with the environment.

N. Chitadze (✉)
Black Sea Region Geopolitical Research Center, Tbilisi, Georgia
e-mail: nikachita@yahoo.com

O. Nikolov, S. Veeravalli (eds.), *Implications of Climate Change and Disasters on Military Activities*, NATO Science for Peace and Security Series C: Environmental Security, DOI 10.1007/978-94-024-1071-6_4

Ecologic problem: the critical ecologic regions of the world. The ecologic situation in the world is approaching to the crisis level, because the demands of the society are not appropriate to the ecologic resources of the biosphere.

The tensed ecologic situation is caused by the several factors; among of them the long and uncontrolled industrial development, the rapid rate of the urbanization process, and the militarization of the environment should be pointed out.

One of the clear examples of the irrational consumption of the nature is the destruction of the forest regions. About 1000 years ago, the huge spaces of our planet were covered by the forest, and the area was 6,2 billion hectares. Today, this area is decreased to 4,2 billion hectares by 1/3. The most intensive exploitation of the forest resources was going on in Europe. If earlier, the forest had about 70–80% of the territory of Europe; today it covers only 30%. During the last period, the area of the tropical forests, where it is concentrated more than half of our flora and fauna, is decreased by 20–30%.

By rapid rates occurs the process of the degradation of the soil. Each year, from the agricultural process, about 6–7 million hectares of the fruitful land have been deducted, which is transferred to the deserted and uninhabited territory [1].

One of the main reasons of the ecologic system's degradation is pollution of the environment by industrial and nonindustrial waste materials. Each year, on the earth about 100 billion tons of the natural materials is processed; from this volume in practice, only 10%, 9/10 of rubbish, is consumed. Industrial waste materials pollute water, soils, and atmosphere. As a result, in the beginning of the twenty-first century, over 1 billion people in the world lacked the access to safe water, and some 2,4 million people – more than a third of the world's population – lacked access to proper sanitation [2].

Due to the abovementioned reasons, it has been appeared the global environmental threat – the noncontrolled change of the air structure – which is interrelated in the violation of the ozone layer of the atmosphere, which creates the threat not only for the lives of the modern but also for the future generations.

The danger, related to the changing of the air's structure, is caused by increasing the level of CO_2 (carbon dioxide) and other gases in the structure of the atmosphere, which is the main reason of the air temperature increase. For example, the carbon dioxide levels in the atmosphere have been increased from 0.027 to 0.036%. At the same time, it is going on the process of decreasing the level of oxygen in the air. For example, each year about 9 billion tons of fuel is burned, for which is spent about 16 billion tons of oxygen. The auto transport annually consumes 5 billion tons of oxygen. If in 1890, the average global temperature of the air was 14.5°, today it has reached 15.2°. Due to it, there has been an increased number of droughts in the USA, China, and post-Soviet space. The increasing of the air temperature has already been reflected on the decreasing of the glacial cover of arctic and Antarctic, also mountain's regions. In case of the continuation of such tendencies and increasing the level of the world ocean, flooding threatens about 3% of the land, where lives about thousand millions of inhabitants [3].

According to the experts, the risks, related to the global warming problem, will be increased in the twenty-first century. For example, within each 40-year period, the temperature will increase for 1°, which means that, during the twenty-first century, the temperature will be increased for 2,5°. This level of the temperature's changing approximately for the 50 times prevails the level of the climate changes during the past 10,000-year period [4].

According to the prognosis of the Organization of Economic Cooperation and Development, as a result of the climate changes and air pollution, till the 2060, the number of victims will be about 200 million people.

Due to the air pollution, annually about 6–9 million people will die, when in 2010 this data was three million.

The direct effect of the pollution can prevail 1% of the World GDP, which is about 2,6 trillion USD per year. With regard to the economic effect of the early death, the costs will be estimated from 18 to 25 trillion dollars, which is equal to the US GDP (5).

Effect of the Climate Changes to South Caucasus Region

All those abovementioned climatic changes on the global level reflected to the situation in the South Caucasus Region. Among the different types of problems, irrational consuming of the mineral resources, pollution of atmosphere and water resources, etc., should be mentioned.

If in the 70–80 years of the last century, the main sources of the air pollution in South Caucasus were industry, especially in the big cities and industrial centers, the level of this sector of economy has been decreased at the end of the twentieth century and was substituted by transport.

According to the specialists, from the capital of Georgia Tbilisi, the atmosphere is erupted about 276,500 tons of the hazardous substances. With regard to the capital of Azerbaijan Baku, the exhaust from the auto transport functioning prevails 349, 2000 tons [6].

Conclusions and Recommendations

Taking into account the main global problems, related to the climate change, it should be taken into consideration many factors, based on the principles of the United Nations Framework Convention on Climate Change, and other important documents, having been adopted on the international level. Among of the various steps, the most important is paying an attention to decreasing the consumption of electricity using the alternative sources of energy, decreasing the using of auto transport and increasing the movement by bicycles, and widening the territory of the protected areas, on the share of which is coming (as to 2010) 10 and 15% of the world's land surface area [8].

One of the key roles in the resolution of the different problems, having been caused by climate and environmental changes, can play NATO. After the ending of the "Cold War," the alliance has developed a network of regional partnership frameworks with 41 partner countries from the Euro-Atlantic area, the Mediterranean and the Gulf region, as well as bilateral relations with other partners from the different regions of the world. NATO acknowledges that it faces many environmental challenges. In this regard, the alliance is working to reduce the environmental effects of military activities and to respond to security challenges emanating from the environment; first of all it concerns the rescue operations.[1]

It is important to mention that the alliance started working over resolving the problems of natural environmental challenges before the foundation of UN program UNEP (United Nations Environmental Program was founded in 1972). By the initiative of NATO, in 1969, the Committee on the Challenges of Modern Society (CCMS) was established. Until its merger with the NATO Science for Peace and Security (SPS) program in 2006, the CCMS created the convenient base for the promoting dialogue between NATO and its partner countries to share knowledge and experience on social, health, and environmental matters, both in the civilian and military sectors.[2]

References

1. Neidze V (2004) Human geography. Edition Lega, Tbilisi, pp 231–232
2. Basic facts about the United Nations. UN Department of Public Information (2004). New York, p 144
3. Elizbarashvili N, Davitashvili Z, Berouchashvili N (2012) Georgaphy. Meridiani, Tbilisi
4. Oxford Illustrated Science Encyclopedia (2008) Translated to Georgian. Edition "Gia Karchkhadze", p 72
5. Simon Upton (2016, August 15). Environment Director at the OECD. Air pollution's true costs. The world's opinion page
6. Bliadze M, Chanturia G, Kereselidze D (2012) Geography. Bakur Sulakauri Publishing, Tbilisi, p 206
7. The United Nations Framework Convention on Climate Change (1994)
8. Soutullo A (2010) Extent of the global network of terrestrial protected areas. Conserv Biol 24:362–363
9. http://unfccc.int/essential_background/convention/items/6036.php
10. http://www.oecd.org/environment/indicators-modelling-outlooks/the-economic-consequences-of-outdoor-air-pollution-9789264257474-en.htm
11. http://onlinelibrary.wiley.com/doi/10.1111/j.1523-1739.2010.01465.x/abstract
12. http://www.nato.int/cps/en/natohq/topics_91048.htm?
13. http://www.nato.int/cps/en/natohq/topics_84336.htm

[1] http://www.nato.int/cps/en/natohq/topics_84336.htm

[2] http://www.nato.int/cps/en/natohq/topics_91048.htm?

Disasters and Security: Key Concepts

Carl Bruch

Abstract This document defines and contextualizes key concepts linking environmental, human, and theater security, including "security," "environmental security," and "human security." It then introduces disaster risk management as a framework for engaging military and civilian authorities in developing a whole-of-government approach and building resilience to climate change and climate-induced disasters.

Keywords Conceptual frameworks • Environmental security • Human security • Disaster risk management

Presentation Overview and Objectives

This document defines and contextualizes key concepts linking environmental, human, and theater security, including "security," "environmental security," and "human security." It then introduces disaster risk management as a framework for engaging military and civilian authorities in developing a whole-of-government approach and building resilience to climate change and climate-induced disasters.

Key Takeaways

There are several ways to define "security." A typical, popular definition is "The quality or state of being secure: as freedom from danger; freedom from fear or anxiety" [3]. A more military-oriented definition is "Measures taken by a military unit, activity, or installation to protect itself against all acts designed to, or which may, impair its effectiveness. A condition that results from the establishment and maintenance of protective measures that ensure a state of inviolability from hostile acts or influences" [9].

C. Bruch (✉)
Director of International Programs, Environmental Law Institute, Washington, DC, USA
e-mail: bruch@eli.org

O. Nikolov, S. Veeravalli (eds.), *Implications of Climate Change and Disasters on Military Activities*, NATO Science for Peace and Security Series C: Environmental Security, DOI 10.1007/978-94-024-1071-6_5

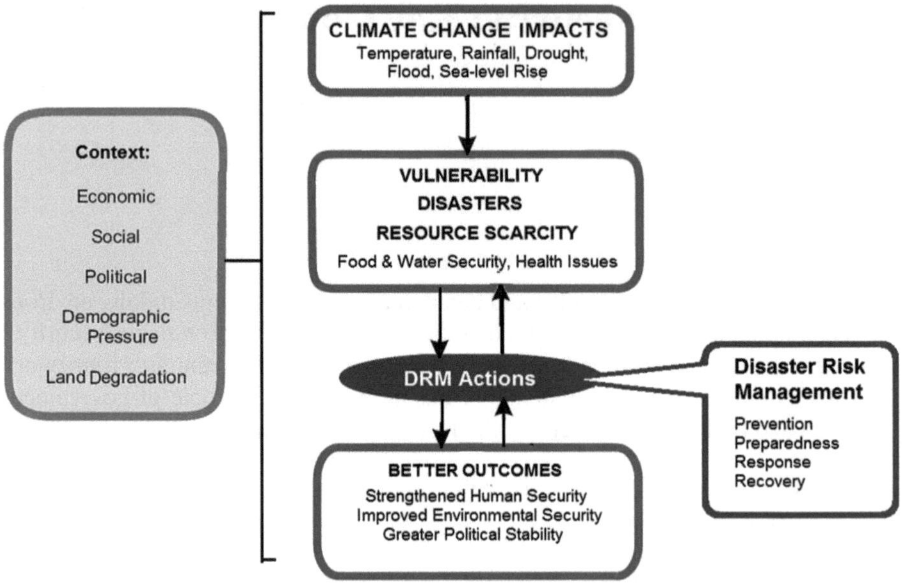

Fig. 1 Conceptual framework for the role of disaster risk management (Adapted, with substantial changes, from UNEP [5])

In the 1980s, and especially in the 1990s, security analysts started identifying linkages between the environment and security, noting that environmental degradation and shortages of potable water and other key resources could contribute to insecurity and the onset of conflict [2, 4]. This was the start of the concept of "environmental security." A common definition of the term is "The protection of important ecosystem services and assurance of a supply of natural resources, including water, soil, energy, and minerals, in order to enable continued economic and social well-being" [1].

Human security is a more recent concept: "Human security is a comprehensive framework for addressing widespread and cross-cutting threats. Recognizing that threats to individuals and communities vary considerably across and within countries, and at different points in time, the application of human security calls for an assessment of human insecurities that is people-centered, comprehensive, context-specific, and prevention-oriented" (UN OCHA 2012). A more eloquent and less operational definition from the UN General Assembly is "The right of people to live in freedom and dignity, free from poverty and despair. All individuals, in particular vulnerable people, are entitled to freedom from fear and freedom from want, with an equal opportunity to enjoy all their rights and fully develop their human potential" [6].

Disaster risk management increases the resilience of societies by improving environmental security and human security. [Resilience is "The ability of a system, community or society exposed to hazards to resist, absorb, accommodate to and recover from the effects of a hazard in a timely and efficient manner, including through the preservation and restoration of its essential basic structures and functions" [7].] Figure 1 illustrates how various climate change impacts can drive vulnerability to disasters and resource scarcity, whose ultimate effects (whether on migration, political instability, or conflict or in improved human and environmental security) depend substantially on disaster risk management.

Disaster risk management has four key components:

- Prevention (the outright avoidance of adverse impacts of hazards and related disasters) and mitigation (the lessening or limitation of the adverse impacts of hazards and related disasters), often through planning such as contingency planning, land-use planning, and pre-disaster recovery planning
- Preparedness (to effectively anticipate, respond to, and recover from the impacts of hazard events)
- Response (provision of emergency services and public assistance during or immediately after a disaster in order to save lives, reduce health impacts, ensure public safety, and meet basic subsistence needs)
- Recovery (the restoration, and improvement where appropriate, of facilities, livelihoods, and living conditions of disaster-affected communities, including efforts to reduce disaster risk factors)

A key aspect of recovery is "building back better," which seeks to reduce future risks and prevent conflict. It includes both "hard" aspects (such as improved infrastructure and equipment) and "soft" aspects (such as improved institution building, capacity building, and training).

Presenter Background

Carl Bruch is the Director of International Programs at the Environmental Law Institute. He has helped countries and organizations throughout Africa, the Americas, Asia, and Europe develop and strengthen their environmental laws, improve institutions, and build capacity. His work focuses on environmental peacebuilding (especially after conflict), environmental governance, adaptation, and environmental emergencies.

Suggested Key Readings

See References.

Conclusion

The linkages between environmental security, human security, and theater security are now well recognized. The relatively recent evolution of disaster risk management provides an important conceptual and operational framework for addressing these linkages in a proactive way. Both civilian and military institutions have important roles in the different phases of the disaster risk management cycle, which will be discussed in further detail throughout this workshop.

References

1. Fiksel J, Hecht A (2012) Environment and security. http://www.environmentandsecurity.org/view/article/167611/. Accessed 14 Dec 2014
2. Homer-Dixon TF (1991) On the threshold: environmental changes as causes of acute conflict. Int Secur 16(2):76–116
3. Merriam-Webster(2014) Security. http://www.merriam-webster.com/dictionary/security. Accessed 14 Dec 2014
4. Myers N (1986) The environmental dimension to security issues. Environmentalist 6:251–257
5. UNEP (UN Environment Programme) (2011) Livelihood security: climate change, migration and conflict in the Sahel. UNEP, Nairobi
6. UNGA (UN General Assembly) (2012) Resolution 66/290. 25 Oct 2012
7. UNISDR (UN Office for Disaster Risk Reduction) (n.d.) Terminology. http://www.unisdr.org/we/inform/terminology
8. UNOCHA (UN Office for the Coordination of Humanitarian Affairs) (2014) Human security unit, strategic plan 2014–2017. https://docs.unocha.org/sites/dms/HSU/HSU%20Strategic%20Plan%202014-2017%20Web%20Version.pdf
9. USDOD (U.S. Department of Defense) (2016) Department of Defense Dictionary of Military and Associated Terms. Adopted 8 Nov 2010, amended 15 Feb 2016. JP-1-02. Available at: http://www.dtic.mil/doctrine/new_pubs/jp1_02.pdf

Disaster Risk Management: General Peer Review of the Disaster Management System in the Republic of Bulgaria

Georgi Petrov

Abstract In the last decades, the negative impact of the natural disasters on human life, economy and environment has increased. Different tools to address this negative trend are made available to the governments, including such assessing the established disaster management systems and proposing concrete and specific recommendations for further improvements of the policies related to the disaster protection. Such a tool is the so-called peer review.

Keywords Peer review • Disaster management • Whole of government approach

General Peer Review of the Disaster Management System in the Republic of Bulgaria

Peer review is a governance tool where the performance in disaster risk management of one country called "reviewed country" is examined on an equal basis by experts called "reviewing peers" from other countries. The process is based on exchange of experiences and results in nonbinding recommendations aimed at policy improvements in disaster risk management and civil protection. The peer review process provides an effective way to:

1. Facilitate exchange of good practices
2. Strengthen mutual learning and common understanding
3. Deliver credible and trusted recommendations

G. Petrov (✉)
Critical Infrastructure, Prevention and Control Directorate to the Directorate-General Fire Safety and Civil Protection, Ministry of Interior, Sofia, Republic of Bulgaria
e-mail: gzpetrov.160@mvr.bg; goshopetrov@gmail.com

O. Nikolov, S. Veeravalli (eds.), *Implications of Climate Change and Disasters on Military Activities*, NATO Science for Peace and Security Series C: Environmental Security, DOI 10.1007/978-94-024-1071-6_6

The objectives of the peer reviews are in particular to:

1. Contribute to improved policy making on national disaster risk management and civil protection through mutual learning and external assessment by reviewing experts from other countries acting as peers
2. Contribute to the development and implementation of relevant EU policies and steer progress in priority actions for the EU cooperation on disaster risk management and civil protection, including where relevant a contribution to the implementation at national level of the international framework for disaster risk reduction (Hyogo Framework for Action)
3. Increase the consistency between the different national disaster risk management and civil protection policies and stimulate transferability of good and innovative practices
4. Foster policy dialogue in Europe and enhance regional cooperation between countries exposed to common or similar hazards and risks
5. Encourage awareness raising through involvement of all stakeholders in the review process and wide dissemination of the results
6. Ensure visibility and political commitment at high level to promote the disaster risk management agenda

The peer review is based on the following principles which are essential for its success: voluntary character, adequate level of commitment, participatory and multi-stakeholder approach, value sharing, mutual trust, coherence, comparability and flexibility, independence, and credibility.

There are two types of peer reviews – general and thematic ones. *The general framework* builds on the post-2015 priorities under the Hyogo Framework for Action and, where relevant, makes links with other relevant international targets/principles (e.g., targets under Sustainable Development Goals). Cross border/international cooperation to strengthen civil protection and disaster risk management is also considered.

The reviewing frameworks for thematic peer reviews cover three specific priorities of the EU disaster risk management/civil protection policy, notably:

Risk assessment, taking into account the 2010 EU guidelines of risk assessment and the framework used for the cross-country analysis of national risk assessments (EU-OECD project). Relevant links with the use of risk assessments in other EU policies is also made (e.g., ex ante conditionality for the use of EU structural and investment funds, sectoral risk assessments (e.g., floods, nuclear hazards, critical infrastructure, climate vulnerability risk assessments, etc.), research projects, and risk mapping).

Risk management capabilities, taking into account the methodology proposed by the Commission in Guidelines for the assessment of risk management capabilities adopted in 2015.

Preparedness strategies, taking into account a range of actions that may be undertaken, encompassing, but not limited to, training programs aimed at civil protection personnel, civil protection exercises, early warning systems, and other technologies, as well as raising public awareness [1].

In 2014, the Republic of Bulgaria was affected by various disasters (storms, heavy rainfalls, hails, floods, and landslides) which substantially damaged public infrastructure and facilities in the field of energy, water resources, telecommunications, transport, public health, cultural heritage, industrial enterprises, and agriculture and also caused loss of human life. According to the statistics for the damages and losses in the report of Germanwatch (called Climate Risk Index), Bulgaria is ranked sixth and in the ranking with the most affected European countries, for the same year, respectively, third, after Serbia and Bosna and Herzegovina. In the report is pointed out that the Climate Risk Index is actually an indicator that measures the exposure and vulnerability to extreme events and should be taken into account by the countries in order to be better prepared for more frequent and intensive disasters in the future [3].

Disasters during 2014 and in particular their negative consequences made Bulgaria to contact the European Commission in order to express its interest in undergoing a general peer review of the disaster management system. As part of its response to the disasters and mainly the floods, Bulgaria decided to reform its disaster management system and to use the peer review's recommendations which to feed into the planned reforms.

The Peer Review Mission Was Conducted over 10 Days from 22 June Until 1 July 2015

During this 10-day mission, the peer review team met and interviewed more than 100 stakeholders from many different organizations (central, regional, and local governmental authorities and agencies, NGOs, academia, and media). By bringing together stakeholders with a variety of backgrounds, expertise and responsibilities, the peer review sessions helped to be achieved one of the key objectives of the peer review process, namely, to share knowledge and encourage cooperation between disaster risk management stakeholders in Bulgaria. The review centered on disaster risk management principles and policies developed at global level (the Hyogo Framework for Action, the Sendai Framework for Disaster Risk Reduction) and at European level (the EU Civil Protection Mechanism and EU directives relating to specific risks). The general review framework covered the following five broad areas:

- Institutional and legal framework
- Risk assessment
- Risk management planning
- Preparedness, response and recovery
- Public awareness

The peer review resulted in a report which clearly points out the strengths of the current system established in the Republic of Bulgaria and gives clear recommendations for further improvements. The report is available on the website of the peer reviews (www.eupeerreviews.eu) both in English and Bulgarian languages. The direct link to the reports is the following: (http://www.falck.nl/nl/peerreviews/publications/articles/peer-review-report-bulgaria-2) [2].

Conclusion

As a follow-up to the final report of the peer review and its recommendations, Act on Amendments and Supplements to the Disaster Protection Act was elaborated and adopted in July 2016 which aims at:

- Shifting from a response-focused emergency management system to a more holistic disaster risk management system.
- Establishing Disaster Risk Reduction Councils to the Council of Ministers, Regional Governors and the Mayors. Those councils shall act as disaster risk reduction platforms at the respective administrative levels.
- Strengthening the role of the National Disaster Risk Reduction Platform through regulating the horizontal and vertical relations at the different levels of governance.
- Putting in place disaster risk reduction planning at national, regional and local levels.
- Optimizing the planning of the effective use of the financial resources.
- Promoting and encouraging both public and private sectors to undertake actions to address the disaster risk management activities.

References

1. Guidelines Peer Review – programme for peer reviews in the framework of EU cooperation on civil protection and disaster risk management 2015–2016
2. Peer Review Report Bulgaria (2016) Programme for peer reviews in the framework of EU cooperation on civil protection and disaster risk management 2016
3. Global Climate Risk Index (2016) Who suffers most from extreme weather events? Weather-related loss events in 2014 and 1995 to 2014, Sönke Kreft, David Eckstein, Lukas Dorsch & Livia Fischer – Germanwatch, 2016

On the Necessity of Improving the Research Infrastructure in the Western Black Sea for the Purposes of Flood Risk Management

Lyubka Pashova, Anna Kortcheva, and Vasko Galabov

Abstract This paper aims to emphasize the necessity to improve the Bulgarian research infrastructure in the western part of the Black Sea region, which can enhance the capability for more effective flood hazard assessment and risk management in the coastal area. To fulfill the requirements of the Directive 2007/60/EC flood hazard and risk maps of the Areas with Potential Significant Flood Risk (APSFR) for coastal zone have to be prepared. Short overview of the approaches for compiling such maps has been presented as well as all parameters' evaluations needed and their uncertainties based on the available information and models. Still existing obstacles to the provision of timely geospatial information from monitoring stations along the coastal area are discussed, and some suggestions for improving the research infrastructure in the western part of the Bulgarian Black Sea coast are listed.

Keywords Flood directive 2007/60/EC • Flood hazard and risk maps • Bulgarian coastal zone • Black Sea

Introduction

One of the key areas for new discussions of the United Nations running initiative Millennium Development Goals is the raising public and institutional awareness for natural and man-made disastrous events ([16]; www.un.org/millenniumgoals). It is widely recognized that most frequent natural hazards across Europe are the floods, whose number and frequency during the last decades are significantly growing due to climate change (e.g., [1]). According to statistics in the last 20 years, flooding has

L. Pashova (✉)
National Institute of Geophysics, Geodesy and Geography – Bulgarian Academy of Sciences, Sofia, Bulgaria
e-mail: lpashova.niggg@gmail.com

A. Kortcheva • V. Galabov
National Institute of Meteorology and Hydrology – Bulgarian Academy of Sciences, Sofia, Bulgaria
e-mail: anna.kortcheva@meteo.bg; vasko.galabov@gmail.com

O. Nikolov, S. Veeravalli (eds.), *Implications of Climate Change and Disasters on Military Activities*, NATO Science for Peace and Security Series C: Environmental Security, DOI 10.1007/978-94-024-1071-6_7

been the most common natural disaster, which is accounted of 43% of all recorded events; as between 1980 and 2010, 3563 distinct flood phenomena have been evidenced in 37 European countries. The research institutions, among the different stakeholders with an interest in the social and legal aspects of flood risk management, disaster risk reduction, and climate change adaptation, contribute substantially to reducing the vulnerability of population, increasing the people's resiliency, and recovering from floods by the expert knowledge and experience in the decision-making process [33]. In 2007, Directive 2007/60/EC, known as Floods Directive [12], created a pan-European framework that can support Member States in identifying, evaluating, and addressing flood risk as complement to the 2000 Water Framework Directive [10].

According to the Water Act and other EU water-related directives transposed in national legislation, the Bulgarian Ministry of Environment and Water has adopted methodologies for preliminary flood risk assessment and for flood hazard and flood risk assessment (http://www.moew.government.bg/?show=top&cid=67). Four river basin management districts have announced on their web pages river basin management plans and integrated flood risk management (FRM) plans for public consultation and written feedback, starting at December 30, 2015, under the requirements of Directive 2007/60/EC. Each FRM plan considers all aspects of risk management, focusing on prevention, protection, and preparedness, including flood forecasts, early warning systems, and the characteristics of each of the river basins for the next 6 years – from 2016 to 2021 inclusive. FRM plan includes a program of specific measures or combination of measures solving the problems and achieving the objectives set for each of the designated areas with potential significant flood risk (APSFR). The analysis presented in this paper concerns mainly issues related to coastal flood risk management and relevant information needed for preparation of flood hazard maps for the Bulgarian coastal zone, which belongs to the basin region of Black Sea Directorate. A short review of the published flood hazard and risk maps for APSFR according to the adopted national methodology is presented, and uncertainties in all parameters' evaluations based on the available knowledge, ambiguities among expert opinions, and their distinct approaches are also discussed. Main difficulties, those still exist in ensuring of accurate and timely geospatial information from available monitoring stations along the coast zone, are outlined. The approach, elaborated in the methodology for drawing flood hazard maps, which have to couple the information on the total water level components (i.e., wind and wave setup, run-up, surge, tide, sea level rise) with terrain characteristics of the coastal area and low-lying land, is briefly outlined. To ensure the needed observational data sets for numerical modeling and significant progress toward modernization of the next coastal flood hazard and risk maps, some suggestions for improving the research infrastructure in the western part of the Bulgarian Black Sea coast are listed.

Flood Events: Most Frequent Natural Disasters

Among the natural disasters, the floods are prevailing part of the natural hazardous events that can cause major material losses and lead to loss of life and injuries. The statistics of the floods in Bulgaria that has gathered by different governing institutions usually is an incomplete and inaccurate, and the analysis of socioeconomic impacts and damages, in most cases, lacks or is rather descriptive. Collected information for the history of past floods is sporadic and is not representative due to lack of definite organizational structure or institution responsible for that purpose.

Frequency and intensity of floods in EU and Bulgaria have clearly increased in the last years mostly caused by heavy rain events (e.g., [2, 28]). Severe floods over the Bulgarian territory are observed in 1997, 2001, 2002, 2005, 2006, and 2014. On Fig. 1 some statistical data for floods that happened in 2000–2015 period are shown. It is worth to mention the riverine floods in 2014, which are total 360 with big

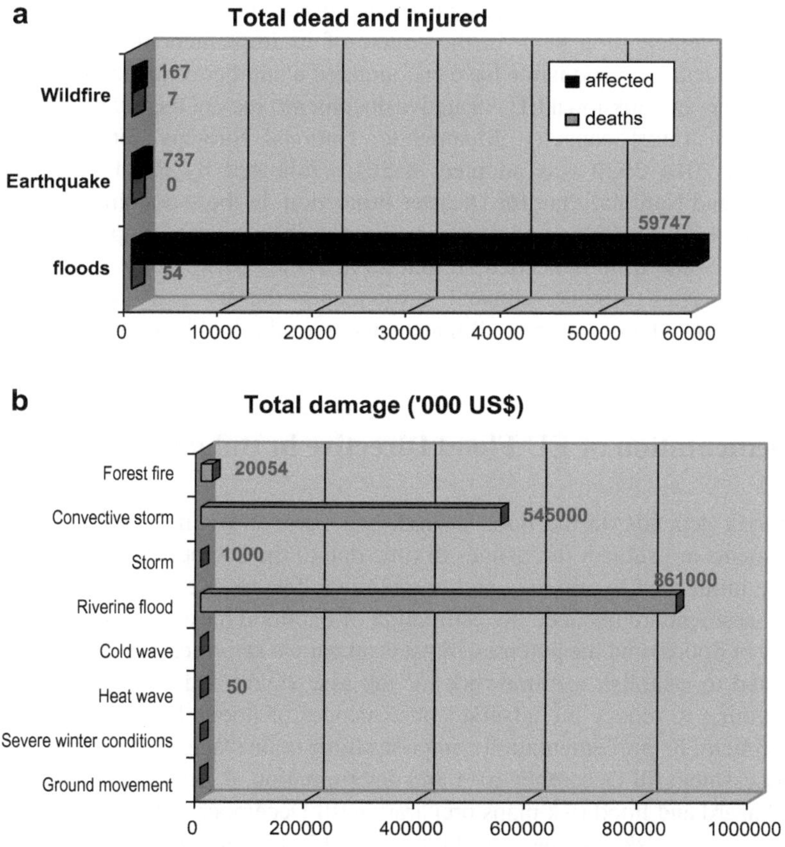

Fig. 1 Natural disasters in Bulgaria/2000–2015/: (**a**) victims and injured; (**b**) economic damages in US$ (Source: Refs. [7, 21, 22])

number of such events in the districts/oblasti/Smolyan, Plovdiv, Montana, Varna, Sofia, Veliko Tarnovo, and Burgas. More than 8500 people have been injured and 20 died and the material damages have been evaluated of about 177,604K BGN [7, 21, 22]. The data are based on annual reports submitted by 141 permanent municipal commissions for protection of population against natural disasters, accidents, and catastrophes.

During the last years, all these circumstances impose permanently amendment of the Bulgarian legislation and its conformity to the European laws and directives. Specific assessment of the flood hazard and risk in the coastal area of Bulgaria has not been carried out before the transposition of FD into national legislation. In this respect, based on scientific studies and expert knowledge, a purposeful state policy to coordinate efforts of interested parties in the field of flood risk management was undertaken. The engineering activities in the coastal zone have been carried out following the normative documents from 80 years of the past century. In contrast to many EU and other countries in the world with gained experience in the coastal flood prevention and the flood risk mitigation, our country, for the first time, had to develop national methodologies, recommendations, and manuals for each stage of the FD implementation [11]. In the course of its implementation, the Bulgarian institutions legally responsible have encountered a number of difficulties and barriers, e.g., new and unknown EU normative documents, lack of the financial resources, and administrative capacity. Meanwhile, National Strategy for Disaster Risk Reduction 2014–2020 was adopted in 2014, followed by Road Map, National Program, and National Plan for Disaster Protection; in these documents key activities along with deadlines and structures responsible for their implementation are described. Some time is needed so that all activities between the institutions and interested parties to be coordinated; subsequently the necessary actions to implement the adopted legislation at all management levels shall be undertaken.

Implementation of EU Flood Directive in Bulgaria

Flood risk depends on the flood hazard, exposure, and vulnerability. These three components are subject to changes in time due to the socioeconomic development and are influenced by climatic and non-climatic factors. Key components of flood risk assessment are the accurate estimation of the flood hazard (magnitude and frequency of floods) and the potential impacts on human activities. EU Floods Directive requested to establish a framework for the assessment and management of flood risks aiming to reduce the adverse consequences of floods for the society and the environment. Its implementation comprises three main stages: (1) preliminary flood risk assessment till December 2011 and determination of APSFR till June 2012, (2) flood hazard and flood risk maps preparation till December 2013, and (3) establishment of Flood Risk Management Plans by 2015 [12]. Second and third stages are subjected to revision and updating every 6 years. During the implementation of FD, some principles are followed: principle of upstream-downstream solidarity, subsidiarity, and collaboration between all public policies and bodies. FRM plan aims to suggest strategic vision for the ARSFR and focusing on the prevention, preparation, protection, and recovery after flood event. It highlights the hazards and risks from rivers, the sea, surface water, groundwater, and reservoirs and set out how risk management authorities will work together with communities to manage flood risk. FRM plans set out where and how to manage flood risk providing most benefit to communities and the environment. Their main purpose is to implement the national priorities, to prioritize measures, and to allocate public budget on efficient and urgent actions to reduce the flood risk.

Recently, following the recommendations of the national methodology, the preliminary flood risk analysis for the region of Black Sea Basin Directorate (BSBD) shows that the number of past events in riverine basins and coastal zone is 761 [27]. The significance of each registered flood event has been evaluated according to four criteria: "human health," "economic activity," "environment," and "cultural and historical heritage." On the territory of the Black Sea Basin, 241 areas with floods have been assessed and 241 cases for which there is likelihood for potential future floods with significant adverse consequences. There are local low-lying areas along the Bulgarian coast of the Black Sea, which are prone to flooding, e.g., Balchik, Varna, Nesebar, Burgas, Primorsko, and Ahtopol. The risk levels have been determined by using the highest values for each area according to nationally approved indicators and criteria. The allocation of fixed APSFR in coastal zone with sea floods is 11 regions with total length of 267 km, which is 25% of all identified APSFRs.

The second stage in the FD's risk management process was the production of flood hazard maps and flood risk maps for the areas identified as APSFR. For that purpose, in 2013 the Bulgarian Ministry of Environment and Water endorsed a national methodology for flood hazard and flood risk assessment [23] elaborated by the National Institute of Meteorology and Hydrology at the Bulgarian Academy of Science. Its Part 3 contains methodological guidelines for hazard assessment and risk caused by sea floods along the Bulgarian Black Sea coast, which was prepared with participation of two of the authors of this paper.

Flood Hazard in the Bulgarian Coastal Zone

Threat of flooding is a conditional probability phenomenon which threatens the life and health of the population in coastal areas, threatening the environment, and can lead to property damage, including the economic loss, damage to the housing, infrastructure, industrial facilities, loss of territory, destruction of cultural and historical heritage, and other possible negative consequences. It is important to note that coastal (or sea) flooding is different from river flooding, which is caused mainly by heavy rainfall. Coastal flooding is generally caused by a combination of high water levels during storm surges with high waves, which lead to overtopping of the coastal defenses and inundation of the low-lying areas. They potentially threaten the human

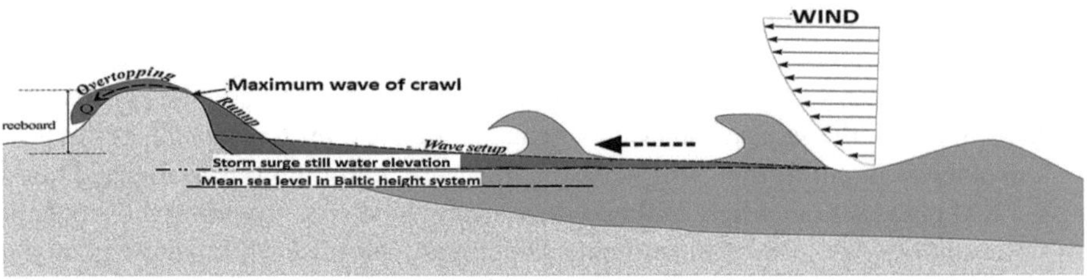

Fig. 2 Scheme of raising the water level by the action of wind waves and storm surge

life and could cause property damages. A storm surge is temporary rises in water level caused by severe weather conditions.

To determine the potentially flooded areas, the sea level increasing should be determined as a result of sharp decreasing in atmospheric pressure in the center of the cyclone and the prolonged effect of the strong wind blowing toward the coast. Sea waves contribute to coastal inundation hazards by three consecutive processes (see Fig. 2, [14]): (1) wave setup, a local elevation in the mean sea level on the foreshore, caused by the reduction in wave height through the surf zone; (2) wave run-up, sum of the wave setup and the wave swash, which is the extra height that broken waves reach as they run up the beach and adjacent coastal barrier (natural or artificial); and (3) overtopping, the spillover of waves as they reach the crest of the coastal barrier or defense structure, resulting in flooding of the land and properties behind the barrier. In some cases, due to the high waves, the existing engineering facilities along the coast such as coastal protection dikes, dams, coast retaining walls, or natural barriers (sand dunes, beach ridges, etc.) are partly or totally demolished. The water masses that enter behind these barriers can lead to further sea level rise and flooding of the territory.

There are two main sources of sea floods: (1) the major source is the rise of the static water level, depending on weather conditions associated with the trend of rising sea level and climate change, including globally and (2) coastal flood is generated by the wind and waves under the stormy conditions or during the storm events. The sea level variation due to tides in the Black Sea is small and reaches ~10 cm [26]; the tides are a deterministic in nature and do not hide uncertainties. Coastal floods began in the contact zone of the sea and land and are caused by powerful high waves, repetitive and attacking the coast, and all the natural and artificial obstacles near the coastline. The greatest damages and destructions, including washing the soil and its deposition into the sea, are observed in the contact zone that is within 100 m inland, where the impact of breaking waves is most significant [13, 14].

Despite the existing hydrometeorological hazards along the western coast of the Black Sea, systematic surveys of inundations due to storm surges and high waves at the Bulgarian coast have not been performed. Available data and information have a narrative character and focus on the description of coastal inundations and their effects due to storm surge events in combination or not with riverine/estuary floods [24, 27]. Important elements of a scientific analysis of stormy situations or severe weather conditions in the western Black Sea are found in some works [4, 20, 29, 30, 32]. Bulgarian researchers from institutes of the Bulgarian Academy of Sciences have participated in the EU projects (e.g., http://www.micore.eu/; http://www.theseusproject.eu/; http://www.increo-fp7.eu/; http://www.risckit.eu/) related with coastal risk assessment, developing modern tools for reliable predictions of the morphological impact of marine storm events in support of civil protection mitigation strategies, and EO-based solutions contributing to an improved preparedness and mitigation planning for areas highly vulnerable to natural disasters and already noticeable climate change trends.

A sufficiently long-term history of accurate measurements of extreme storm winds, waves, and storm surges along the Bulgarian Black Sea coastal zone is usually rare to be found. In the absence of measured data covering the whole coastline in old periods, the hindcasting storm surge and waves can satisfy the need for historical data [4, 18, 19]. Historical archives and databases of NIMH and the Geocartfund of the Agency of Geodesy, Cartography, and Cadastre provide the initial information about available observations of the sea level and waves, which were then carefully analyzed and refined with human expertise in order to select the severe storm situations along the Bulgarian coastal zone. The synoptic situations associated with these events accompanied by coastal hazards are also examined. The final data set includes ten cases within the period 1976–2012 corresponding to the most relevant well-documented storm conditions associated with high waves and/or high storm surges over the Bulgarian coast of the Black Sea. As an example, the estimated storm surge water level by the storm surge model forced by the downscaled ERA-Interim historical reanalyses of 10 m wind and mean sea level pressure along the Bulgarian coast of the Black Sea during the storm in 7–8 February 2012 is shown in Fig. 3 (EU project IncREO, http://www.increo-fp7.eu/).

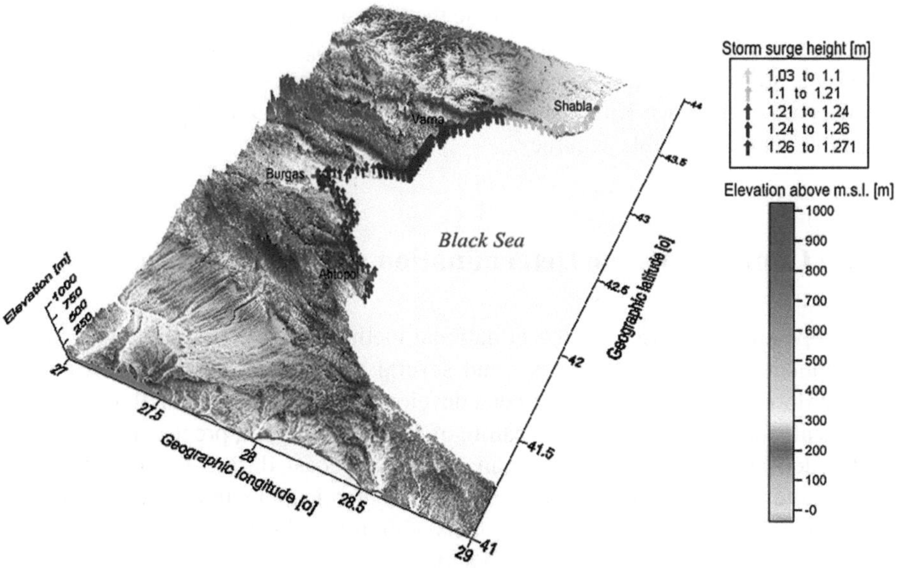

Fig. 3 Estimated storm surge height on 7–8 February 2012 along the Bulgarian Black Sea coast

Coastal Flood Hazard and Flood Risk Assessment: Bulgarian Methodological Guidance

In the nationally adopted methodology, coastal flooding in the Bulgarian Black Sea coast is regarded as a special group [23]. Based on the requirements of Article 6 of Directive 2007/60/EC, Article 146e and Article 146f of the Water Act, flood hazard maps, and flood risk maps shall be developed for all identified APSFRs. The maps should be prepared for the following cases:

- Floods with a low probability, where the likely return period is larger than or equal to 1000 years, as well as extreme event scenarios
- Floods with a medium probability, where the likely return period is larger than or equal to 100 years
- Floods with a high probability, where the likely return period is larger than or equal to 20 years, where appropriate

Due to the specific physical processes along the coastal zone, the determination of the maximum water levels with certain probability is accompanied by uncertainties and further complicating the calculation and modeling of coastal floods due to lack of the necessary data sets. For example, the coastal area around the town Pomorie is one of the potentially most vulnerable zones for sea flooding, where each of the six observed sea floods for the period 1999–2010 has caused significant material damage. The guidelines for determining the maximum sea levels under various scenarios and variants are developed taking into account the basic physical processes and phenomena related to the coastal flood risk. Based on PEFRBSBD [27], collected information and expert knowledge, the methodical guidance contains consecutive steps of actions to determine the coastal flood hazard by calculating the maximum sea levels with different probability according to the FD recommendations. The basic steps for drafting the flood hazard and risk maps have been demonstrated for the area of the town Kiten, since only for this area a digital elevation model was available with reasonable accuracy.

Main Components for Determination of Total Sea Level

In Europe and worldwide, different national methodologies and guidelines, practical applications, "good practices," and several approaches for the assessment of coastal flood hazard and risk have been developed. It should be noted that due to the nonlinear stochastic and dynamic nature of coastal processes, precise determination of flooded areas along the Bulgarian Black Sea coast during strong sea storms implies the use of advanced specialized hydrodynamic and wind wave models. The development or adaptation of such models with licensed software for 3D modeling, however, requires significant financial and human resources, special education, and training of qualified specialists to work with them as well as requires powerful

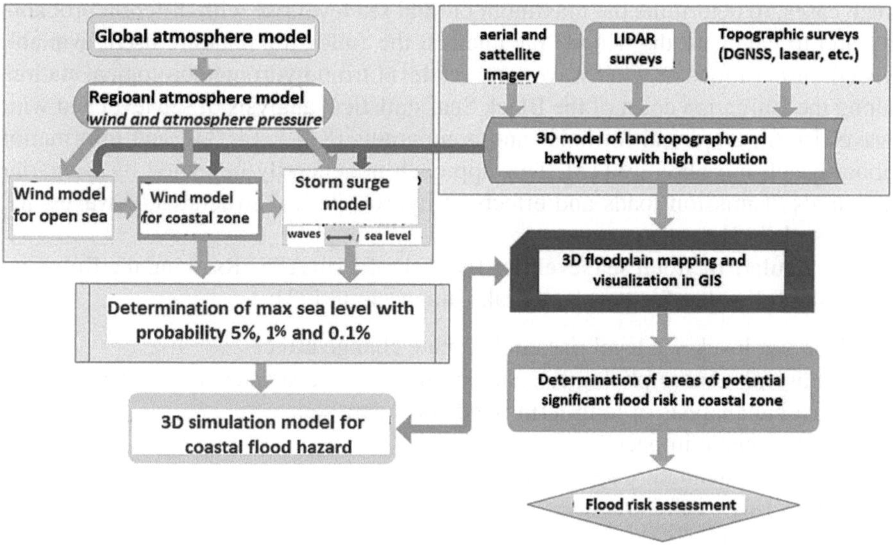

Fig. 4 General scheme for coastal flood hazard and risk assessment

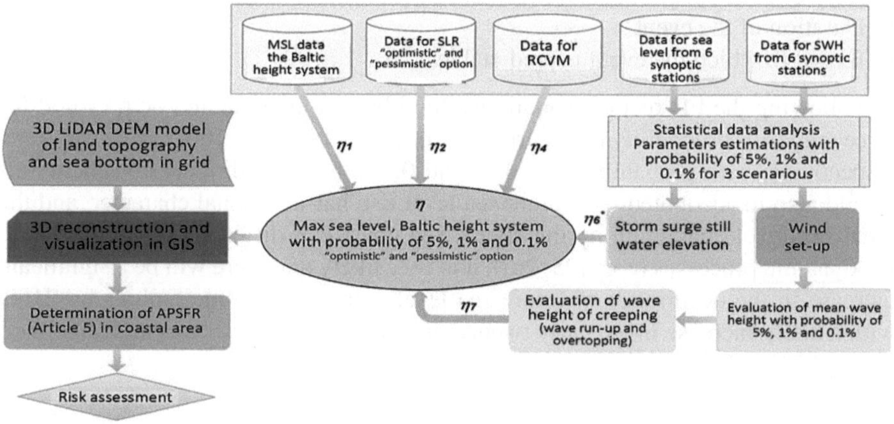

Fig. 5 Scheme of the sequence of actions that apply in the methodological guidelines for evaluating the threat and the risk of sea flooding (part 3: methodological guidance, [23])

hardware platforms for processing large data sets. In this case, the scheme of successive steps for coastal flood risk assessment, shown on Fig. 4, can be followed [23]. Otherwise, when necessary highly specialized software programs are lacking, the scheme presented on Fig. 5 can be used. In the methodological guidance, the present and future hydrometeorological hazards are determined by combined effect of strong winds, high waves and storm surges in the changing climate conditions, the geological peculiarities of coastal zone, and the anthropogenic influence. In such cases, to determine the maximum or total sea level rise with different probability and to calculate the needed parameters, the following data are used: available visual observations of wind waves and sea level from hydrometeorological stations along the Bulgarian coast of the Black Sea, statistical analysis of sea level and wind waves [5, 6], empirical formulas and nomograms [8, 13, 14, 34], and information about beach morphology [17]. This approach is generally regulated by Bulgarian standards "Emission loads and effects of hydrotechnical structures of waves, ice, and vessels" [3].

To calculate the total sea level rise for each of 11th APSFRs along the Bulgarian coast, the following factors should take into account [23]:

- Mean sea level, sea level rise, and climate change effect
- Geological, geomorphological, and oceanographic characteristics of coastal areas
- Statistical analysis of long-term wind wave data sets
- Anthropogenic impact
- Changes in the nature and dynamics of coastal processes
- Engineering infrastructures, barriers, artificial and natural walls, and embankments
- Spatial extent
- Indirect effects
- Intensity of impact in the coastal zone
- Duration of the event
- Extent to which the event is predictable

Following the FD recommendations, flood hazard maps should be prepared for three ranges of probability of occurrence – once per every 20, 100, and 1,000 years, respectively, with probability 5%, 1%, and 0.1%. Some additional considerations should also to accounted: the future sea level rise has a regional character, and the local sea level change along the Black Sea coast can differ significantly due to geodynamic processes (e.g., [25, 26]); it is very likely that there will be a significant increase in the occurrence of future sea level extremes in some regions by 2100, with a likely increase in the early twenty-first century [1].

Data and Models for Flood Maps Production

For the purposes of the developed methodological guidelines for assessing the sea flooding hazard and risk, an analysis of available information and databases is performed [23]. The values for individual components forming rising the sea level with different probability of occurrence can be identified by: (1) a statistical analysis of data from long-term measurements/observations, (2) by calculating a theoretical formulas or numerical models, and (3) by combining both methods – through calculation based on theoretical models and usage the results of statistical processing of some parameters determined in the coastal meteorological and geodetic monitoring stations. The data sets for local key parameters have been looked for to be accurate, homogeneous, standardized, and with good spatial coverage; DEM of land and seabed, sea level, wave, wind, geological data, river discharge, geomorphological, and oceanographic features of coastal areas; and runoff, erosion and abrasion, sea currents, infrastructure, and facilities built in the coastal zone. Hydrometeorological, geophysical, and oceanographic parameters data can be obtained in two ways: (1) long-term observations/measurements and (2) as a result of numerical simulation of atmospheric, storm surge, and wave models with high resolution. The total sea level is determined following the scheme on Fig. 5.

Bulgarian methodology for flood hazard and risk assessment uses a combined approach, which integrates data from long-term observations of different hydrometeorological parameters and statistical analysis. These data includes: wave parameters/ significant wave height/from six coastal weather stations from the monitoring system of the National Institute of Meteorology and Hydrology – BAS, sea levels, and wind data. To determine the height of wave creep on coastal swath, nomograms of the American Guide for coast protection are used (Shore Protection Manual 1984). Methodological guidelines recommend usage of empirical formulas for determining the wave run-up when the coastal slope cannot be used directly in the nomograms [14, 15]. If input data for the geometry of barriers are available, empirical formulas to determine the amounts of water overtopping the barriers can be used. Due to engineering facilities along the Bulgarian coast of the Black Sea, the exact calculation of wave propagation in wave creep in coastal slope is hampered. In the methodological guidance the maximum sea levels at wave creep in coastal slope, the overtopping of water mass over obstacles and moving water behind the barrier is recommended to perform by averaging the wave height of creep. A similar approach was used and described in the literature (e.g., Coastal Floodplain Mapping – Guidelines and Specifications: Final Report, KWL File No.2785.001, [9]). This approach of action is applicable for the coastal flood risk assessment based on the available information at that time of the methodological guidelines developing. The coastal flood hazard evaluation is performed following the basic steps of a simplified algorithm. The results of statistical processing of data from meteorological observations at synoptic stations of the NIMH-BAS, data analysis of measurements on the Black Sea level, and research results related to climate change (two scenarios for future sea level rise – optimistic and pessimistic) are used. The determination of possible inundated areas from sea flooding with three probabilities 5%, 1%, and 0.1% and two scenarios according to Article 146 of WA for the case study of the town Kiten is done with ArcGIS software tools, and different flood hazard and risk map have been drown [23].

Improving Scientific Infrastructure in the Western Black Sea for Flood Risk Management

The methodological guidelines for coastal flood hazard and risk assessment contain some uncertainties and inaccuracies. Some of them are due to the following factors: (i) limited resource of qualitative data of measured parameters of the state of coastal systems and processes because of poorly developed research infrastructure, (ii) the use of probability distributions of the theory of extreme events to assess the extreme sea level rise due to storm surge, (iii) limited knowledge of the future development of coastal systems, (iv) difficulty of predicting the socioeconomic development of coastal areas, etc. Methodological guidelines for flood hazard and flood risk assessment comply with existing data sets and the results of statistical processing of the characteristics of sea level and wind wave parameters along the Black Sea coast as well as geomorphological features of coastal zone. Further studies on quantification of the couplings between sea state (mean and extremes) and flooding hazards today and ahead are needed to assess the exposure, vulnerability, and flood risk on APSFRs along the western Black Sea coast.

To overcome some difficulties, obstacles regarding needed database of observations and to use new generation numerical models for the next phase of Directive 2007/60/E implementation, a strengthening and improvement of scientific infrastructure in the western Black Sea for flood risk management is desirable. The coastal flood hazard and risk mapping can be significantly improved if the following scientific and technological issues can be resolved:

- Refinement of the numerical hydrodynamic and wave models
- Evaluation of SLR and determination of uncertainties regarding magnitude, location, and temporal occurrence of the storm events
- Effective method of geo-visualization
- Establishing of National Spatial Data Infrastructures according to EU INSPIRE Directive
- Developing web-based emergency management and early warning systems
- Developing web-based geospatial decision support tool and applications – Google Maps, OpenStreetMap, etc.

Additionally, social economic issues that could contribute to the improvements of compiled flood hazard and risk maps after the next 6-year period are:

- Preparing coastal flood hazard and risk maps according to end users' requirements
- Efficiently spreading of information among the stakeholders and the general public of flood hazard and flood risk to improve the awareness of coastal flood risks and already undertaken measures
- Consultation and collaboration with all "interested party"

To improve flood resilience to extreme coastal events at national level, the following purposeful activities could be foreseen:

- Establishing co-located meteorological, hydrological, geodetic, and oceanographic monitoring stations for long-term monitoring of the air-sea-land interaction processes in real time.
- Optimization and modernization of observational infrastructure to strength the capacity to provide early warnings of extreme events.
- Improving research and administrative capacity.

- Shift the focus toward prevention, preparedness, and mitigation measures to the storm surge risk reduction.
- Improve preparedness of the stakeholders and general public for disaster risk mitigation in the Bulgarian Black Sea coastal region.

Conclusive Remarks and Recommendations

Coastal storm hazards are the result of a complex interplay of wind, seawater, and local coastal features. The best tools for assessing these hazards are numerical models combined with statistical data analysis. Wave and storm surge numerical models can be used for the long-term hindcast as well as for issuing coastal hazard and flood warnings. Some recommendations for the future work related to the flood risk management of the Bulgarian coastal zone can be suggested:

- Further development and improvement the current methodology for the preparation of flood hazard and flood risk maps
- Development of a coastal flood hazard forecasting method for early warnings
- Raising the public awareness to the topic of coastal flood risk management and disaster risk mitigation
- Developing a comprehensive database of past and recent flood events and their impacts in the coastal zone
- Improvements of the monitoring and exchange of information among all stakeholders on national and international level
- Climate change to be considered when developing national strategies, programs, plans, and other documents for the integrated coastal zone management and spatial planning
- Improvement of the impact assessments of a variety of environmental parameters and their changes and human activities in coastal areas
- Use of the results of interdisciplinary research as basis for the formation of policies for sustainable development of the coastal regions

The present methodology has already been applied for Bulgarian coast of the Black Sea for the preparation of flood hazard and flood risk maps of APSFRs, which are logical continuation of the preliminary flood hazard and risks assessment [27] and the main part of the Flood Risk Management Plans. Two examples of elaborated flood hazard maps for pessimistic and optimistic scenarios respectively are shown on Figs. 6 and 7 for the region of Varna, which have been prepared according to the methodological guidance provided in [23].

A proper evaluation of available data on present sea level change and storm surge events is vital as the outcome forms the basis for a number of important socioeconomic decisions and developing the Bulgarian policy for flood risk prevention and reduction. The concept of coastal flood risk management could shift from defense against floods to a more integrated approach. Evaluation of the coastal flood hazard and risk as well future sea level rise applying various climate scenarios simulations

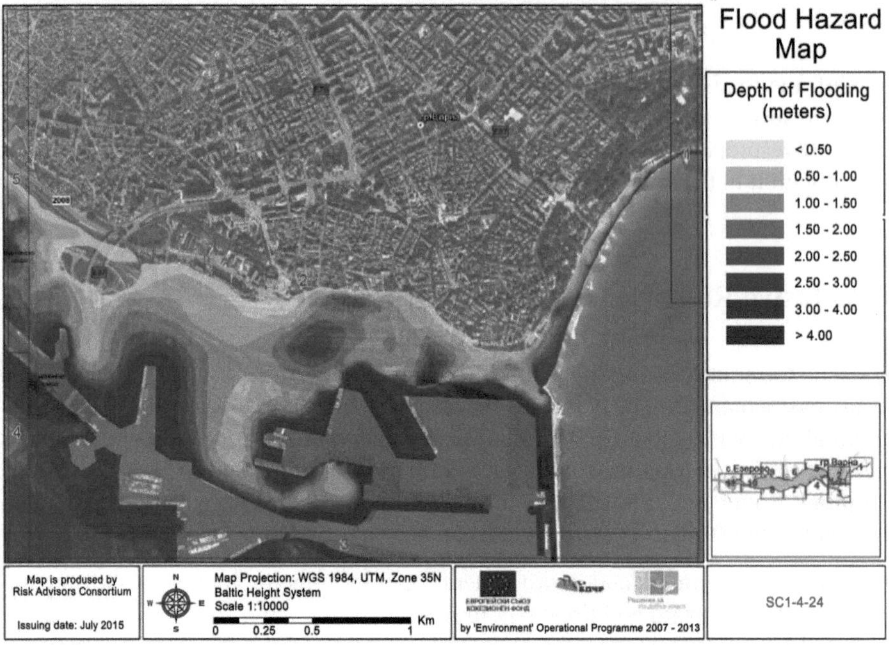

Fig. 6 Flood hazard map for optimistic scenarios for Varna bay (Source: Web page of Black Sea Basin Directorate: M04_24_HazardmapDepth_BS_04_1000_02_SC1.PDF)

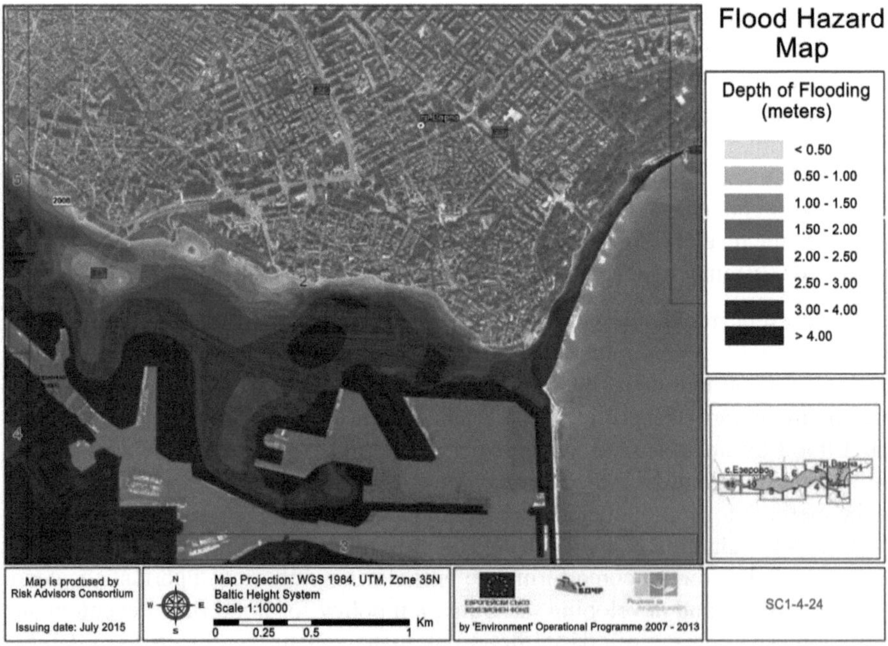

Fig. 7 Flood hazard map for pessimistic scenarios for Varna bay (Source: Web page of Black Sea Basin Directorate: M04_24_HazardmapDepth_BS_04_1000_02_SC1.PDF)

for this century should be based on the knowledge-based approach. The interdisciplinary researches contribute to increase capacity for coastal flood risk assessment, better prediction of storm surges, and effective coastal flood risk management.

References

1. Church JA, Clark PU, Cazenave A, Gregory JM, Jevrejeva S, Levermann A, Merrifield MA, Milne GA, Nerem RS, Nunn PD, Payne AJ, Pfeffer WT, Stammer D, Unnikrishnan AS (2013) Sea level change. In: Stocker TF, Qin D, Plattner G-K, Tignor M, Allen SK, Boschung J, Nauels A, Xia Y, Bex V, Midgley PM (eds) Climate change 2013: the physical science basis, Contribution of Working Group I to the Fifth Assessment Report of the Intergovernmental Panel on Climate Change. Cambridge University Press, Cambridge, UK
2. Dakova, S. (2006) Recent extraordinary floods in Bulgaria, observing and modelling exceptional floods and rainfalls. In: Proc. of the AMHY-FRIEND International Workshop on Hydrological Extremes, University of Calabria, Cosenza. 3–4 May 2006, 8p
3. Emission loads and effects of hydrotechnical structures of waves, ice, and vessels (1988) "Legal basis of design and construction" – special edition of KTSU, 1988
4. Galabov V, Kortcheva A, Bogatchev A, Tsenova B (2015) Investigation of the hydrometeorological hazards along the Bulgarian coast of the Black Sea by reconstructions of historical storms. J Environ Prot Ecol 16(3):1005–1015
5. Grozdev, D. (2005) Wind wave climate along the Bulgarian Black Sea coast. PhD thesis, Institute of Geography, Institute of Oceanology, BAS, Sofia, 183 p. (in Bulgarian)
6. Grozdev, D. (2008) Extreme characteristics of the Black Sea level along the Bulgarian coastline. In: Proceedings of national conference with international participation GESCIENCES'2008, BGS, Sofia, 125–126 (in Bulgarian)
7. Guha-Sapir, D., R. Below, PH. Hoyois (2016) EM-DAT: the CRED/OFDA International Disaster Database – www.emdat.be – Université Catholique de Louvain–Brussels–Belgium, http://www.emdat.be/
8. Coastal Engineering Manual (2002) Part II, Publ. Number EM 1110-2-1100, CECW-EW, 30 April, Department of the Army, US Army Corps of Engineers, Washington
9. Coastal floodplain mapping-guidelines and specifications: final report, KWL File No. 2785.001, 2011
10. COM 120 (2015) The water framework directive and the floods directive: actions towards the 'good status' of EU water and to reduce flood risks, COM (2015) 120 final, Brussels, 9.3.2015
11. EC (2016) EU overview of methodologies used in preparation of flood hazard and flood risk maps, Final report, 118pp. doi:10.2779/204606
12. FD (2007) Directive 2007/60/EC of the European Parliament and of the Council of 23 October 2007 on the assessment and management of flood risks, OJ L 288, 6/11/2007
13. FEMA (2004). Final draft guidelines for coastal flood hazard analysis and mapping for the Pacific Coast of the United States, 344pp., https://www.fema.gov/coastal-flood-hazard-analysis-and-mapping
14. FEMA (2005) Coastal flood hazard analysis and mapping guidelines wave runup and overtopping. Focused Study Report, 56p., http://www.fema.gov/media-library-data/20130726-1541-20490-9494/frm_p1wave2.pdf
15. FEMA (2008) Guidance for coastal flood hazard analyses and mapping in sheltered waters. Technical Memorandum, Washington, DC, 22p
16. ISSC (2015) Review of the sustainable development goals: the science perspective. International Council for Science (ICSU), Paris. 92pp. http://www.icsu.org/publications/reports-andreviews/review-of-targets-for-the-sustainable-development-goals-the-science-perspective-2015/SDG-Report.pdf

17. Keremedchiev, S. (2001) Morphohydrographic analysis of the coastal zone in the Bulgarian Black Sea littoral, Proc. of Institute of Oceanology, Vol.3, Varna, BAS, 57–64 (in Bulgarian)
18. Kortcheva, A., G. Kortchev, V. Galabov (2010) A wave prediction system for real time sea state forecasting in the Black Sea, Bulg J Meteorol Hydrol,vol.15, N2,. 56–66
19. Kortcheva, A., V. Galabov, M. Dimitrova, A. Bogatchev (2014) Hindcast of extreme hydrometeorological events along the Bulgarian Black Sea coast, In: Proc. of Int. Conf. "Analysis and Management of Changing Risks for Natural Hazards", 18–19 Nov 2014, Padua, available at: http://www.changes-itn.eu/Conference/Programme/DetailedProgramme/tabid/157/Default.aspx
20. Mungov, G., A. Kortcheva (1994) Modeling of storm surges in the western Black Sea with application of Operational UK Met Office Data, In: Proc. of the IC "Black Sea 1994"; Varna; 12–14 Sept 1994, 63–64.
21. NSI (2016). Crisis events, http://www.nsi.bg/en/content/13267/crisis-events-occurred (27.05.2016)
22. NSI (2016). Floods, http://www.nsi.bg/en/content/13286/earthquakes.
23. MFHRA (2013) Methodology for flood hazard and flood risk assessment, as required by Directive 2007/60/EU, Ed. D. Dimitrov, NIMH – BAS, Sofia, available at: http://www.moew.government.bg/?show=top&cid=67 (in Bulgarian).
24. Palazov A, Stanchev H (2009) Risks for the population along the Bulgarian Black Sea coast from flooding caused by extreme rise of sea level. Inf Secur Int J 24:65–75
25. Pashova, L. (2008) The Black Sea level as an indicator of climate and global change, International conference "Global environmental change: Challenges to science and society in Southeastern Europe", Conference CD proceedings, 19–21 May 2008, Sofia, 6 p
26. Pashova L (2012) Assessment of the sea level change on different timescales from Varna and Burgas tide gauge data. C R Acad Bulg Sci65(2):193–202Proceedings of the BAS
27. PEFRBSBD (2012) Preliminary flood risk assessment in the Black Sea Basin Directorate, MEW, BSBD, Varna, 117 p., available at http://www.bsbd.org/bg/index_bg_2934486.html (in Bulgarian)
28. Santato S, Bender S, Schaller M (2013) The European floods directive and opportunities offered by land use planning, CSC Report 12. Climate Service Center, Germany. 79p
29. Surkova G, Arkhipkin V, Kislov V (2013) Atmospheric circulation and storm events in the Black Sea and Caspian Sea. Cent Eur J Geosci 5(4):548–559
30. Trifonov V, Trifonova L (1988) Typification of synoptic situations initiating high waves along Black Sea coast, Probl Geogr 2:42–47 (in Bulgarian)
31. Trifonova, E., N. Valchev, St. Keremedchiev et al. (2014) Mitigating flood and erosion risk using sediment management for a tourist city: Varna, Bulgaria. In: Zanuttigh B, Nicholls RJ, Vanderlinden J-P, Burchart HF, Thompson RC (eds) Coastal risk management in a changing climate. Butterworth-Heinemann, pp 358–384
32. Valchev N, Trifonova EV, Andreeva NK (2012) Past and recent trends in the Western Black Sea storminess. Nat Hazards Earth Syst Sci 12(4):961–977. doi:10.5194/nhess-12-961-2012
33. United Nations Inter-Agency Secretariat of the International Strategy for Disaster Reduction (UN/ISDR) (2004) Living with risk, a global review of disaster reduction initiatives. United Nations, Geneva. 2004
34. US Army Corps of Engineering (2002) Coastal engineering manual. EM 1110-2-1100

Floods in the Republic of Serbia (2014): Response and Lessons Learned

Katarina Strbac

Abstract The Republic of Serbia was affected by floods in May 2014. The level of immediate security threat imposed an obligation to the national government to use its armed forces, within a comprehensive effort to control the disaster, prevent even greater damage and protect human lives. Engagement of the army was significant and in several elements crucial for the success of the entire operation.

Keywords Serbia • Flood • Emergency • Crisis management • Response

Introduction

In the third week of May 2014, Serbia was hit by floods,[1] just as the two of its neighbouring countries (Croatia and Bosnia and Herzegovina).[2] The disaster came as a consequence of exceptionally heavy rains which were caused by the low atmospheric pressure system over the Adriatic region. The recorded weekly amounts of rainfall in Western Serbia within the critical period were equivalent to 3 months of rain under usual conditions. Officially, it was the highest quantity of the rainfall recorded in the region, since systemic measurement has been started (120 years ago).[3]

[1] *Serbia Floods 2014*, United Nations Serbia, European Union and the World Bank Group, Belgrade, 2014: http://ec.europa.eu/enlargement/pdf/press_corner/floods/20140715-serbia-rna-report.pdf

[2] ReliefWeb is a specialized digital service of the United Nations Office for the Coordination of Humanitarian Affairs (OCHA): http://reliefweb.int/disaster/ff-2014-000059-srb

[3] International Sava River Basin Commission (ISRBC): http://savacommission.org/index.php?idnews=148&l=bhbo&page=news_detail

K. Strbac (✉)
Directorate for European Integrations and Project Management, Ministry of Defence, Belgrade, Republic of Serbia
e-mail: katarina.strbac@mod.gov.rs

O. Nikolov, S. Veeravalli (eds.), *Implications of Climate Change and Disasters on Military Activities*, NATO Science for Peace and Security Series C: Environmental Security, DOI 10.1007/978-94-024-1071-6_8

As a consequence of constant raining, water level in the main rivers of western, south-western, central and eastern part of the country[4] was increasing significantly, causing flash floods, widespread floods and landslides. The disaster affected about 1.6 million people living in 38 municipalities and towns. Most of the hit settlements were located in central and Western Serbia. Two cities[5] and 17 municipalities[6] were severely impacted.

The scope of the growing security threats overcame capabilities of the interior security system and local self-government, imposing necessity of the national armed forces to be involved. Fulfilling their constitutional and legal obligations, Serbian Armed Forces provided assistance to civil institutions of the government. The army joined emergency teams at the river banks immediately. It also took control over local crisis management headquarters. Military units supported evacuation of inhabitants, provided temporary shelters and medical care for displaced population, contributed to supplying in food, drinking water and medicaments and had a significant role in the consequence management.

The purpose of this paper is to analyse the legal basis, methodology of operation and experience obtained through the engagement of Serbian Armed Forces in the activities related to a crisis management operation which was performed as a response to a major natural disaster – floods in 2014.

Legal and Strategic Framework for Engagement of the SAF in Civil Emergency Management

Responsibilities of the Serbian Armed Forces, conditions under which they may be used out of the national territory as well as their mandatory subjection to democratic and civil control have been defined by Articles No. 139, 140 and 141 of the National Constitution.[7] According to the Article 139, "The Serbian Armed Forces shall defend the country from external armed threats and perform other missions and tasks, in accordance with the Constitution, Law and principles of international law, which regulate the use of force".[8]

The Law on Defence, within its Article 41, provides a more precise framework of operation, allowing that the units of the Serbian Armed Forces may be engaged in

[4] Rivers: Sava, Tamnava, Kolubara, Jadar, Zapadna Morava, Velika Morava, Mlava and Pek.

[5] Cities of Šabac and Sremska Mitrovica.

[6] Municipalities of Obrenovac, Mali Zvornik, Krupanj, Ljubovija, Vladimirci, Koceljeva, Šid, Svilajnac, Paraćin, Ub, Lajkovac, Ljig, Osečina, Mionica, Smederevska Palanka, Trstenik and Bajina Bašta.

[7] Constitution of the Republic of Serbia, version in English language, official website of the government: http://www.srbija.gov.rs/cinjenice_o_srbiji/ustav_odredbe.php?id=217

[8] Constitution of Serbia, V Organization of Government, Article 139, version in English language, source official website of the government: http://www.srbija.gov.rs/cinjenice_o_srbiji/ustav_odredbe.php?id=222

delivering assistance to the population "in case of natural and other large scale disasters in which lives and health of people and animals, as well as material wealth are endangered in a part of the country's territory, following a request made by the bodies in charge of protection and rescue of people and material wealth".[9]

According to the *Law on the Serbian Armed forces* (Article No. 2), "the Serbian Armed Forces are an organized armed force defending the country from outside armed threats and executing other missions and tasks in accordance with the Constitution, law and principles of the international law regulating the use of force. The President of the Republic or the Defence Minister, authorized by the President can decide that the Serbian Armed Forces should deliver assistance to the competent state body or organization, autonomous provinces body or local self-government body, following their request, in order to protect lives and security of the people and property, provide environmental protection or for any other reasons stipulated by Law".[10]

The Law on Emergency Situations (Article 12)[11] further implies that "in events when all other forces and resources of the protection and rescue system are not sufficient for efficient protection and rescue of people, material and cultural goods and the environment from catastrophes caused by hazards natural and other disasters, the Ministry of Defence shall at the request of the competent department ensure participation of organizational parts of the Ministry of Defence, commands, units and institutions of the Serbian Army to assist in protection and rescue", in accordance with the Law. When the units of the Serbian Armed Forces take part in protection and rescue, they shall be under the command of their senior personnel, in line with the decisions of the emergency management headquarters managing and coordinating protection and rescue activities.

The strategic framework for engagement of Serbian Armed forces in emergency situations is provided by the *National Security Strategy*, *Defence Strategy* and the *SAF Doctrine*.

Chapter 4.3 of the *National Security Strategy* defines that "the Armed Forces of Serbia develop capacities for active contribution to building a favorable security environment, partnership and participation in multinational operations with a UN mandate, as well as for support to civil authorities in countering security threats".[12]

---

[9] *Law On Defence*, Official Gazette, no. 116-07, December 2007: http://www.mod.gov.rs/multimedia/file/staticki_sadrzaj/dokumenta/zakoni/law%20on%20defense.pdf

[10] *Law on the Serbian Armed Forces*, Article 2, Official Gazette, No. 116-07, December 2007. http://www.mod.gov.rs/multimedia/file/staticki_sadrzaj/dokumenta/zakoni/Law%20on%20the%20SAF.pdf

[11] *Law of Emergency situations*, Official Gazette of Serbia, Vol. 111/2009, 92/2011 and 93/2012, version in English language, source website of the South Eastern and Eastern Europe Clearinghouse for the Control of Small Arms and Light Weapons (SEESAC): http://www.seesac.org/res/files/failovi/547.pdf

[12] *National Security Strategy of the Republic of Serbia*, version in English language, IV National Security Policy, 4.3. Defence Policy, Belgrade, 2009, p.p. 26, source official website of Serbian MoD: http://www.mod.gov.rs/multimedia/file/staticki_sadrzaj/dokumenta/strategije/Strategija_nacionalne_bezbednosti_RS_eng.pdf

Missions of the Serbian Armed forces were declared in 2009 by the *Defence Strategy of The Republic of Serbia*, according to which they are: "defence of the Republic of Serbia from armed threats from outside, – participation in building and maintaining peace in the region and the world, and – support to civil authorities in countering security threats".[13] The *Doctrine of Serbian Armed Forces* briefly defines that operations of civil-military cooperation (CIMIC) include cooperation and coordination between civil and military structures in peace, state of emergency and war. Operations of support to civil authorities in resistance to nonmilitary security threats may be performed in case of natural disasters, industrial or other accidents and epidemics.

The *Service Regulation of the Serbian Armed Forces* (Article 170) allows the Minister of Defence to decide (under authorization by the President of the Republic) on engagement of the military in assistance to other public institutions (which may include agencies of the national government, provincial authorities and local self-government), upon their request, in protection of lives, human security and property, environment protection or for other legal reasons. Article 171 of the same document further defines responsibilities of the Chief of General Staff and senior military leadership in response to emergency situations which may be applied only under special authorization of the President of the Republic.

Command and Control System of the Serbian Armed Forces in Emergencies

National command and control system of emergency management in Serbia is structured in three levels (strategic, operational and tactical). *State Headquarter for emergency situation* is the central institution subordinated directly to the government. Autonomous provinces and districts also form provincial or district headquarters. At the bottom of the list, there are city and municipal headquarters for emergency situation management, being established by units of local self-government (cities or municipalities). Units of the police, local self-government and civil defence operate subordinately to the municipal or city headquarters.

For each of the listed civil emergency management organizational units and levels, there is an equivalent level/unit within the military chain of command, authorized to communicate with it and provide assistance in response to disasters. Thus, on the strategic level, the Minister of Defence and Chief of the General Staff of Serbian Armed Forces are members of the State Headquarter for emergency situation. Major organizational units of the Ministry of Defence and the General Staff, such as the

[13] *Defence Strategy of the Republic of Serbia*, version in English language, 5. Defence System of the Republic of Serbia 5.1. Structure of defence system, Belgrade, 2009, p.p. 13, source official website of Serbian MoD: http://www.mod.gov.rs/multimedia/file/staticki_sadrzaj/dokumenta/strategije/Strategija_odbrane_RS_eng.pdf

Table 1 Command and control system in the emergencies

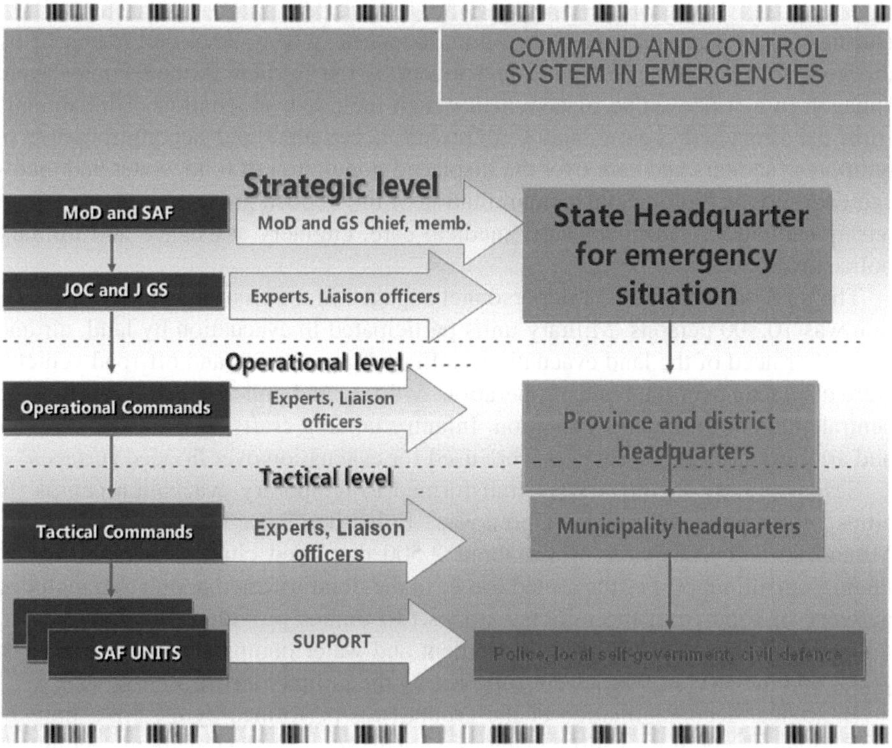

Joint Operations Command and Departments of GS (from J-1 to J-9) may provide experts and liaison officers seconded to the State Headquarter for emergency situation.

Operational commands of Serbian Armed Forces which are regularly subordinated to the Joint Operations Command and Departments of GS cooperate with provincial or district emergency management headquarters, providing liaison officers and experts upon their request and in accordance with available capacities. Tactical commands of SAF (being subordinated to operational commands) are directly linked with municipality headquarters for which they also provide liaison officers and experts. Finally, the units of the Serbian Armed Forces (being at the end of the chain of command and subordinated to tactical commands) cooperate directly with units of the Police, local self-government and civil defence, providing them assistance in the field. The network of communication and cooperation between civil and military institutions in emergency crisis management is presented at image No. 1 (Table 1).

Engagement of the Serbian Armed Forces During the Floods

During the crisis management operation, launched in May 2014 and managed by the government, units, personnel and assets of the Serbian Armed Forces were engaged in a waste scope of activities which included: evacuation of inhabitants from the impacted regions and settlements, acceptance and accommodation in temporary shelters and care over the displaced population of food, water and medicaments supply, biological decontamination of the flooded ground, water pumping, terrain sanitation, veterinary care, medical care, enginery assistance and military police tasks.

The total number of the SAF personnel engaged in the crisis management operation was 10.300 persons. Military units participated in evacuation by land, air and water. For need of the land evacuation, military buses, lorries and off-road vehicles were displaced over the zone of operation. Military and police helicopters were the central element of the air evacuation. Infantry landing crafts, amphibious vehicles and armoured personnel carriers were used for evacuation over flooded surfaces.

Seven military facilities were transformed to temporary evacuation camps (in cities of Sremska Mitrovica, Obrenovac, Belgrade, Pozarevac, Valjevo, Gornji Milanovac and Svilajnac), while about 2.500 displaced citizens were hosted in them. Logistic support of the armed forces to the flood-affected population included delivery of food from the military stocks (20 tonnes provided), drinking water, toiletries, medicaments, logistic equipment and water pumps. About 63 tonnes of goods and 52.000 l of water were supplied by the military airlift.

Biological decontamination of the ground was performed by the NBC units of the Serbian Armed Forces. Fifteen teams were deployed in the theatre to decontaminate 231 acre of terrain and roads, as well as nearly 1.200 buildings.

Military engineers installed 12 heavy mechanized bridges and one floating (pontoon) bridge. Enginery of the SAF was also engaged in dam strengthening, reconstruction and building of roads, landslide recovery, terrain clearing and preparatory construction works.

The military medical support to the civil authorities during the crisis management operation included one officer seconded to the Crisis Management Staff in Obrenovac whose engagement lasted 13 days, three medical teams located in impacted cities and one field hospital deployed in the theatre of operation. The medical team in Belgrade remained active for 9 days. The presence of military medical teams in two other cities was needed longer (Obrenovac 12 days and village of Jakovo, near Obrenovac, 15 days). The field hospital in Obrenovac remained functional for 12 days, providing medical care for 900 patients.

Within the humanitarian assistance to civilian community and environmental protection, the army provided veterinary support to the communities hit by floods (especially flooded villages and farms) through collection of animal carcasses, collection of strays and abandoned animals, collection of groceries destroyed by floods and measures of disinfection, disinsection and rodenticide.

International Assistance

International aid to the Serbian government was provided immediately. Rescue teams and equipment arrived from the Russian Federation, Montenegro, Macedonia and Belorussia. Assistance through the EU civil protection mechanism was provided by Bulgaria, Germany, Romania, Denmark, Slovenia, Hungary, Czech Republic, Austria, France, Croatia and Cyprus. Special rescue teams of the United Nations and the European Union were deployed in the theatre rapidly. The necessary equipment was donated by individual countries, such as Japan, China and Azerbaijan, but also by the international organizations: United Nations, World Food Programme, European Union, OSCE, UN Office for Coordination of Humanitarian Affairs (OCHA) and the United States Agency for International Development (USAID).

Aftermath Activities

After the phase of immediate response to the disaster had ended, assistance of the armed forces to the civilian authorities was continued through terrain sanitation and engagement of the military engineers in the flooded area. Lessons learned were analysed on the strategic level which resulted by concrete tasks given to the command and control level. Through the defence budget and external sources of funding (primarily international aid and donations), material resources were provided for procurement of new equipment (including transport helicopters and equipment for aerial geo-monitoring of high resolution). Plans of operation in emergency situations have been reviewed and updated too. Increased level of attention was paid to development of capability for response to civil emergency situations.

Conclusion

During the coordinated response to the unprecedented floods in 2014 in Serbia, the protection and rescue forces were activated on time. Specialized civil crisis management forces were overstretched, which requested assistance of the military. The available amount of rescue equipment was insufficient due to extraordinary emergency, so the international aid was of great importance. The entire activity requested a complex coordination and performance of comprehensive field activities.

The experience obtained through the performed operations showed that exchange of information performed timely is essential for the success of operations. Aid requests were often sent to inadequate addresses, bypassing some levels of the crisis management structure. For example, in some cases, assistance of the armed forces used to be requested directly from the local level of self-government, before all the resources of provincial or city authorities were fully exploited and without adequate check whether or not in other institutions there were sufficient capacities to solve the issues for which the immediate assistance of the military was requested. Procedures of forces coordination certainly have to be further developed and improved, in order to provide one point to which all the requests are going to be addressed. It is also essential to respect principles of gradual engagement, according to objective estimation of forces and equipment which may be needed for response to disasters in impacted areas. Introduction to practice of the listed measures would create more favourable conditions for deployment of the civil protection mechanism and Serbian Armed Forces in theatre of future crisis management operations.

The lessons learned from the activities during the floods in our country imply that a permanent early warning system needs to be established for greater efficiency in response to future security challenges. Prevention has to be further developed. It is also essential to provide and have always available dedicated forces and resources for rapid reaction. Plans for engagement of additional forces and resources have to be developed. The legal framework for adequate emergency response system has to be adapted. Procedures for acceptance of international aid need to be liberalized so that in emergency situations we do not have cases when equipment which may be necessary in the theatre stand on border posts for many days, waiting for custom or other forms of approval.

Based on the presented analysis, we might propose the prevention plans to be reviewed and implemented after the revision. Mobile teams for command and control in the field should be established too, while more attention should be given to planning and execution of joint exercise.

Natural disasters, including climate ones, are getting stronger and more intensive in the last decade, but the number of victims is kept steady. It is logical to expect that disasters caused by climate changes will probably create new catastrophes in the future. Floods from 2014 were "hundred year floods" – not only because of the level of rainfall but also because of other factors (erosion, deforestation, terrain configuration, etc.). Changed geomorphological structure of the terrain can result in earthquakes all across the region in the future. Having on mind the potential risks, cross-border cooperation seems to be essential for the coming years.

The level of education and training for local authorities has to be increased, as well as the communication on the topics related to emergency situation management in the entire society. In control of flood areas, control and construction monitoring have to be improved. An integrated system for prevention of floods and other disasters in view needs to be developed. Risk analysis has to be performed continually and the EU Flood Directive 2007/60/EC implemented. Prevention and resilience to floods has to be an inseparable part of general development in the Serbian society and the region of Southeastern Europe.

References

1. Serbia Floods (2014) United Nations Serbia, European Union and the World Bank Group, Belgrade, 2014. http://ec.europa.eu/enlargement/pdf/press_corner/floods/20140715-serbia-rna-report.pdf
2. ReliefWeb. Specialized digital service of the United Nations Office for the Coordination of Humanitarian Affairs (OCHA): http://reliefweb.int/disaster/ff-2014-000059-srb
3. International Sava River Basin Commission (ISRBC): http://savacommission.org/index.php?idnews=148&l=bhbo&page=news_detail
4. Republički geodetski zavod: http://www.rgz.gov.rs/template4.asp?PageName=poplave_2014&MenuID=0040122
5. Constitution of the Republic of Serbia. Version in English language, official web site of the Government: http://www.srbija.gov.rs/cinjenice_o_srbiji/ustav_odredbe.php?id=217
6. Law on Defense. Official Gazette, no. 116–07, December 2007.: http://www.mod.gov.rs/multimedia/file/staticki_sadrzaj/dokumenta/zakoni/law%20on%20defense.pdf
7. Law on The Serbian Armed Forces. Article 2, Official Gazette, No. 116-07, December 2007. http://www.mod.gov.rs/multimedia/file/staticki_sadrzaj/dokumenta/zakoni/Law%20on%20the%20SAF.pdf
8. Law of Emergency situations. Official Gazette of Serbia, vol 111/2009, 92/2011 and 93/2012, version in English language, source: web-site of the South Eastern and Eastern Europe Clearinghouse for the Control of Small Arms and Light Weapons (SEESAC): http://www.seesac.org/res/files/failovi/547.pdf
9. National Security Strategy of the Republic of Serbia. Version in English language, IV National Security Policy, 4.3. Defence Policy, Belgrade, 2009, p 26, source: official web-site of Serbian MoD: http://www.mod.gov.rs/multimedia/file/staticki_sadrzaj/dokumenta/strategije/Strategija_nacionalne_bezbednosti_RS_eng.pdf
10. Defence Strategy of the Republic of Serbia. Version in English language, 5. Defence System of the Republic of Serbia 5.1. Structure of defence system, Belgrade, 2009, p 13, source: official web-site of Serbian MoD: http://www.mod.gov.rs/multimedia/file/staticki_sadrzaj/dokumenta/strategije/Strategija_odbrane_RS_eng.pdf
11. Izvestaj o elementarnoj nepogodi – poplavi koja je zadesila Republiku Srbiju i merama koje su preduzete radi spasavanja stanovništva i odbrane ugroženih mesta od poplava, Parliament of the Republic of Serbia, source: official web-site of Serbian Parliament: http://www.parlament.gov.rs/upload/archive/files/lat/pdf/akta_procedura/2014/2220-14Lat.pdf

Military Medical Detachment for Emergency Response: Tool for Disaster Medical Education and Interaction

Rostislav Kostadinov and Alexander Dimitrov

Abstract When disaster strikes there are shortage or scarcity of medical means and capabilities for provision of adequate and rapid medical support to all in need; therefore, all available resources have to be immediately mobilized. In order to maintain the victims' life till medical teams arrive, disaster medical preparedness among population is required.

The severe, extraordinary conditions, developed as a consequence of disastrous events and the urgency of support required for their management reassemble the recoded during wartime demands for medical support.

The aim of this study is to present the capabilities of the Military Medical Detachment for Emergency Response (MMDER) in the disaster medical education and training.

Materials and Methods – By the means of descriptive and comparative methods, the requirements for disaster medical education and training were analyzed and compared with the MMDER capabilities to respond to these requirements. Cluster analysis was performed in order to group some of the most meaningful activities performed or to be planned for achieving better disaster medical readiness among medical specialists, search and rescue teams, and population as a whole.

Based on the performed analyses, the great MMDER capacity for disaster medical theoretical, but mostly practical, training for civilian healthcare providers has to be noted.

Keywords Disaster medical support • Disaster medicine • Disaster medical education • Education and training • Military Medical Detachment for Emergency Response

---

R. Kostadinov (✉) • A. Dimitrov
Bulgarian Military Medical Academy, Sofia, Bulgaria
e-mail: rostikosti@abv.bg

O. Nikolov, S. Veeravalli (eds.), *Implications of Climate Change and Disasters on Military Activities*, NATO Science for Peace and Security Series C: Environmental Security, DOI 10.1007/978-94-024-1071-6_9

Introduction

When disaster strikes the time for planning rescue operations is extremely limited. When discussing the medical assistance to affected by the calamity population, has to be highlighted that any delay into medical aid provision is equal to human lives and abilities lost. Therefore no one is surprised that when disastrous event occurs, the second issue (after the notice about the type, place, and time for the event's occurrence) is about the casualties and what medical activities are undergoing in order to minimize the disaster's impact on human life and health.

By definition disaster could be declared only in case when the required resources for managing the consequences are insufficient in such quantity and quality that this disparity requires urgent and focused external assistance [1]. From this definition we could conclude that when disaster strikes there are shortage or scarcity of medical means and capabilities for provision of adequate and rapid medical support to all in need. Based on this assumption and in accordance with the disaster management doctrine, all available resources have to be immediately mobilized. But from the definition, the specific medical capabilities are scarce and the time for external provision is life costing. One possible solution to this enigma is to prepare as much as possible first disaster medical aid providers such as medical specialists, rescue teams, and population. This wideness of possible responders is dictated from the uncertainty of the calamity – no one could be 100% sure that she or he will be not present in the area of damage and thus included among casualties or survivors. What is more, in the most important for the life saving first 30–60 min after the injury, the medical means and capabilities in the area of damage are limited. Therefore, the casualties and the present into the area arriving for medical assistance medical specialists, will expect support into medical aid provision by everyone available.

These are just few of the rationalities for building centers for disaster medical education and training. Unfortunately there are several other issues that are calling for mandatory disaster medical education among the abovementioned groups.

The severe, extraordinary conditions, developed as a consequence of disastrous events and the urgency of support required for their management reassemble the recoded during wartime demands for medical [2] and other salvation operations. Therefore, the lawmakers implied task to the Bulgarian Armed Forces to be ready to support civilian governmental structures in case of disaster. According to # 67 of the Law for Armed Forces of Republic Bulgaria "in peacetime, in cases of natural disasters, industrial accidents and dangerous environmental pollutions, the Armed Forces are involved in prevention, direct protection of the civilian population, and search and rescue operations." In order to fulfill this, the Bulgarian Armed Forces Medical Service Authorities in the early 1990s proposed creation and establishment of military medical structure specialized in disaster and crisis relief operations and medical support. As a result in 1992 the Military Medical Detachment for Emergence Response (MMDER) was founded on functional principle within the Military Medical Academy. In a decade time after several interventions in different types of calamities, MMDER had been evaluated as a structure of national importance;

therefore on 01 October 2000, the detachment was established as a separate, independent structure, which is part of the Military Medical Academy.

The aim of this study is to present the capabilities of Military Medical Detachment for Emergency Response in the disaster medical education and training.

Materials and Methods – By the means of descriptive and comparative methods, the requirements for disaster medical education and training were analyzed and compared with the MMDER capabilities to respond to these requirements. Cluster analysis was performed in order to group some of the most meaningful activities performed or to be planned for achieving better disaster medical readiness among medical specialists, search and rescue teams, and population as a whole.

Results and Discussion

The first requirement for disaster medical education and training is the recorded trend in disaster frequency and severity. The last three decades have provided us with undoubted evidence regarding steady increase in both the frequency and severity of disastrous events [3]. What is more, the analyses of the causes of this increase revealed that great majority of the factors leading to the observed trend are related to the very essence of our contemporary world development:

1. Industrialization is leading to developing of more sophisticated instruments and apparatuses, as well as industrial process. They are utilizing as sources or have as final products hazardous for human health and life materials – toxic chemicals, ionizing radiation emitters, or/and venomous biological products. Industrialization, in the other hand, is increasing the average workforce income, thus leading to amelioration of the living conditions. This results in increase of the average life expectancy and unprecedented population growth. To summarize the impact of industrialization:

   - it is increasing the hazards,
   - it is increasing the population at risk,
   - it is increasing exposure to these hazards

   As obvious evidence - the changing number of auto vehicles in the last decades is increasing the environmental pollution that negatively impacts human health and endangers human life.
2. With globalization the process of industrialization is widening its geography – the abovementioned changes are recorded almost all over the world.
3. Both the industrialization and globalization are triggering the next contemporary world phenomenon – the urbanization. This enlargement of the cities is not only elevating the health risk due to the increased population at risk (increased population density) but also is creating a lot of obstacles to the medical support in case of disaster:

- Firstly – how to reach the casualties – traffic congestions are common to every modern town, even without any stress and panic among the citizens.
- The multistory buildings are real impediment to survivor's evacuation, hindering access of the rescue and medical teams, as well as to the medical evacuation of the casualties.
- The modern megapolis is an ideal place for area of biological damage development – the slums, so typical for the nowadays cities, are an incubator of biological hazards, the sanitary and hygiene are favourable for easy spreading of the communicable diseases causers, thus creating conditions for epidemic outbreaks.
- Moreover the increasing demands for water, food, and energy delivery for these enormous habitats are leading to environmental exhaustion and increased production – the core of the observed climate change.

4. These three processes are exacerbating the social inequity leading to social unrest in its different forms, some of them classified as disasters: civil war, terrorism, and international armed conflicts – local or regional wars.

Contemporary disaster medical manager is challenged in its planning and organizing by the changes in the medical community. Analyzing the trends into medical environment development into the last decade, the following warnings for disaster medical support execution process are noted:

1. Increase lack of physicians. The young generation is interested in higher payment professional path; therefore, majority of the countries are facing decrease in medical coverage. Even in countries, where the payment of physicians is relatively high (the USA, for instance), the required by growing population number of physicians, is greater than available ones (Fig. 1).
2. Narrow specializations. This phenomenon is very easily explained – physicians and the other medical personnel are income oriented in choosing the specialty of work. This leads to increasing scarcity of medics that are eager to dedicate their lives to the wanted by disaster medical support specialties – disaster medicine, general medicine, internal medicine, general surgery, infectious diseases and epidemiology.
3. Highly dependent on sophisticated equipment. Nowadays the technological development has significant impact on modern medicine. The implementation of the scientific discoveries into medical practice is providing medics with powerful equipment that leads to increasing treatment capabilities. But the negative aspect of this trend is that medical specialists are becoming extremely dependent in their way of thinking, diagnosing, and treating on technological equipment. Unfortunately in greater part of the disasters, this equipment is expected either to be not available or not operational (e.g., disrupted electricity or water distribution); this dependence is leading to decrease in the level of readiness and preparedness of the medics to solve the health threat relying on their knowledge and personal skills.

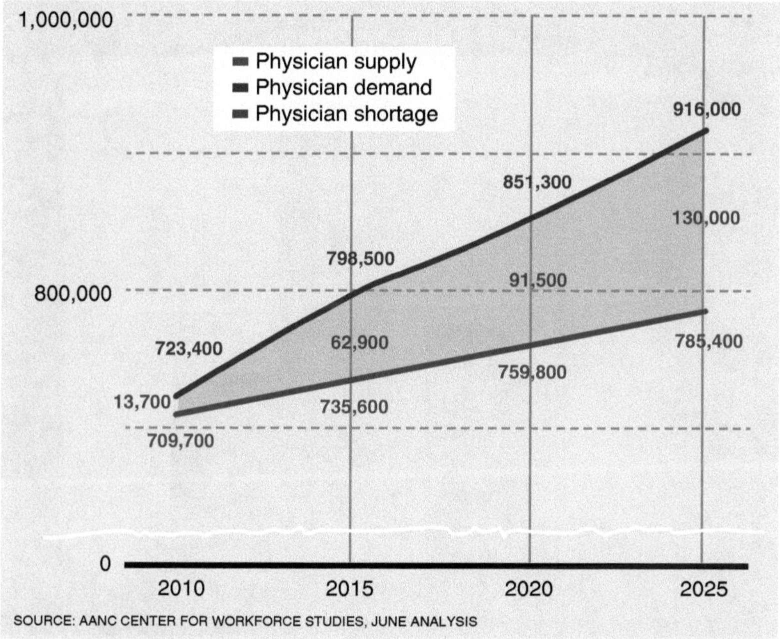

Fig. 1 Growing population physicians

4. The age. The relative or absolute scarcity of physicians was mentioned above, but the analysis on medical community describes one negative trend – the aging. The physicians are not sufficient in number, but even those that are available are above the age, when new knowledge and skills could be easily obtained and when the prolonged duty shifts and work on the field could be easily, physically, and psychologically managed (Fig. 2).
5. Constant physicians' movement. The globalization is providing every single medical specialist with the opportunity to find better proposal for her/his career development. This also has negative impact on disaster medical support readiness, because the disaster medical manager could be surprised by discovering that the physician, who is supposed to lead the triage group, had packed her/his belongings and went abroad 5 days prior to disaster occurrence.

Greater part of disasters are characterized by developing areas of damage involving intoxications, radiation damages, blast or gunshot injuries, terrorist attacks, emerging or reemerging diseases or combination between them, or mass casualty situations. Almost 90–95% of the medical specialists have forgotten the management of such a situations, because their practice is dealing with totally diverse type of patients [4]. And when disaster strikes, it could be expected that knowledge from the university years is not easily recalled. Therefore it is recommendable that these disaster medicine specifics are to be periodically trained.

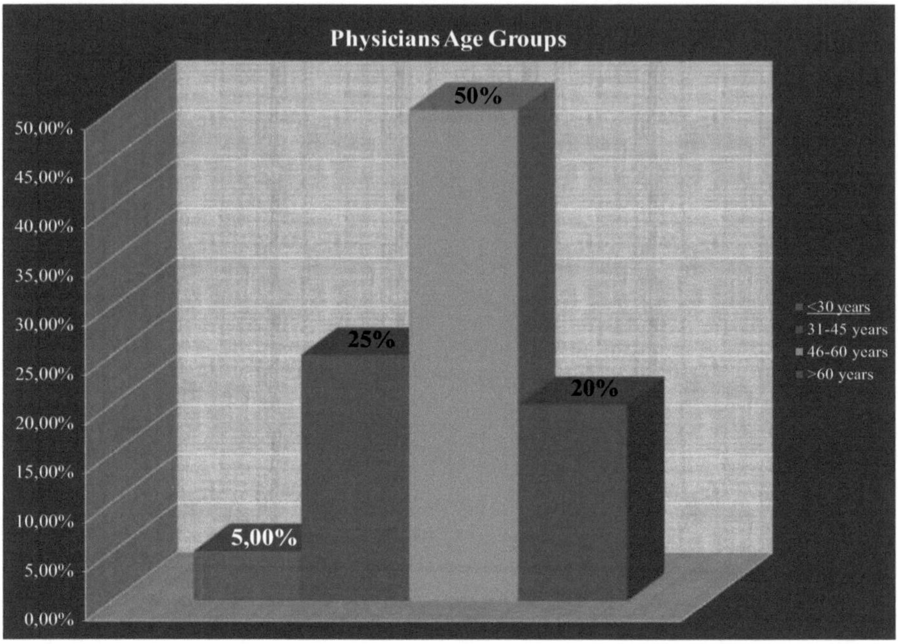

Fig. 2 Physician age groups

Another challenge is the damaged lines of communication, disturbed information sharing, and blocked routes of transportation.

As MMDER is established for supporting civilian healthcare system in case of crises and disasters, its structure is designed to respond to the abovementioned civilian medical community shortfalls (Fig. 3).

When required in accordance with the disaster's damaging factor, the MMDER could in timely manner move close to the affected population and build medical triage-treatment-evacuation capabilities in four different modules:

1. Temporary trauma center
2. Temporary center for treatment of intoxicated patients
3. Temporary center for treatment of patients with acute radiation syndrome or other ionizing radiation injury
4. Field infectious disease hospital

All these modules are staffed with medical personnel that has gained invaluable practical experience during disaster relief missions or by medically supporting military missions abroad.

All the modules are mounted on all-terrain trucks. Containers are used as training base for the MMDER staff monthly training.

Specialized all-terrain and armored ambulances are available for medical evacuation of the stabilized patients.

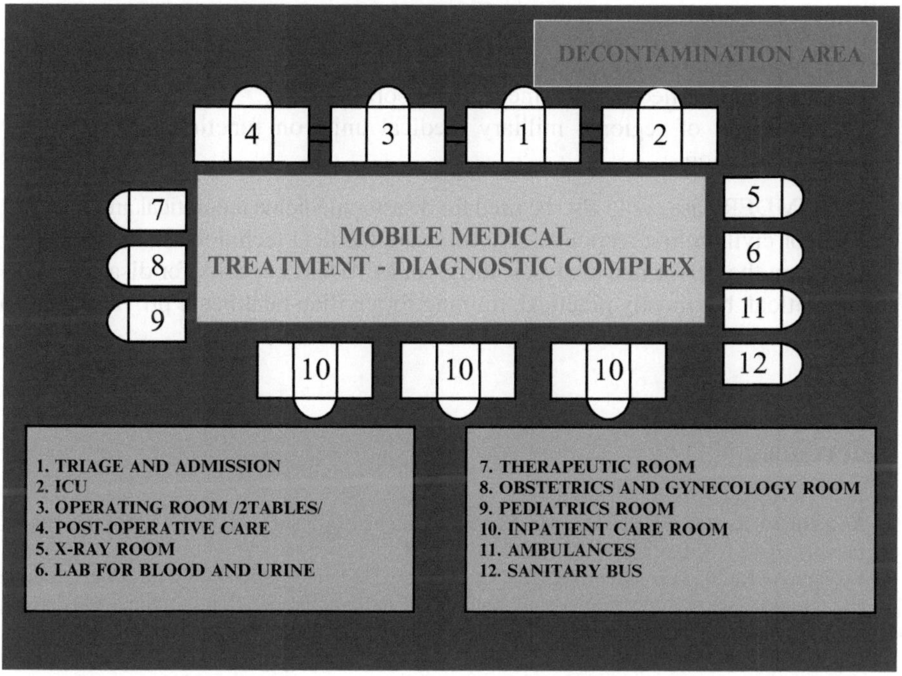

Fig. 3 Medical complex

Military medical staff of the detachment is on 24 h, 7 days a week on-call duty – the only one specialized medical structure on high readiness status in Bulgaria.

The MMDER personnel is working into the clinics of the Military Medical Academy to maintain its medical knowledge and practice on the highest possible level that is attractive for the medics. On a regular basis, a specialized disaster medical training, dealing with diverse mass casualty situation, is performed [5, 6].

Comparing the MMDER capabilities and the civilian healthcare challenges in planning and organizing proper and adequate disaster medical support operation, the following joint educational and training initiatives could be proposed to be implemented in the civilian medical personnel postgraduate education and training:

1. Courses for medical specialists, led by MMDER specialists
2. Disaster medical planning
3. Disaster medical support in case of earthquakes, floods, chemical industrial incident, nuclear plant incident, terrorist attack, and implementation of weapons of mass destruction
4. CBRN medical protection
5. War surgery
6. Field medicine
7. Mass casualty triage
8. Medical evacuation – indications, stabilization, and management during transportation

Joint practical events:

1. Joint field trainings and exercises
2. Common and unified disaster medical support planning
3. Establishment of regional military medical units on functional principal for regional reaction

The MMDER base could also be used for disaster medicine theoretical and practical courses for civilian first responders, nurses, and medical technicians [7].

Based on the performed analyses, the great MMDER capacity for disaster medical theoretical, but mostly practical, training for civilian healthcare providers has to be noted.

References

1. Kostadinov, R. 600 simple steps for disaster medicine exam success. VAP Publishing House, Plovdiv, 2012 г. 176 p. ISBN: 978-954-8326-56-8
2. Kostadinov R (2013) Disaster medicine and military medicine similarities with combat trauma and trauma system//international conference: civil-military cooperation in trauma and combat trauma system. Education and training, 26–27 September 2013. Nunziatella Military School, Naples , p 17Abstract Book
3. https://www.unisdr.org/we/inform/disaster-statistics. .Accessed 10 Aug 2016
4. Noschese, G., R. Kostadinov. CBRN medical teams protection in case of disasters//expert conference civil-military cooperation: enhancing combat trauma system and disaster medical management capacities, 12–14 September 2012, Nunziatella Military School, Naples, Abstract Book, 2012, p. 26
5. Kostadinov, R., A. Dimitrov, K. Kanev. Military medical readiness for chemical and biological terrorists' attacks. Medical management of chemical and biological casualties. Editors Major General Tonev, Kanev, Dishovsky. Irita Publishing House. Sofia, 2009, pp 44–49, ISBN 978-954-9993-91-2
6. Kostadinov R, Dimitrov A (2012) Military Medical Academy, Sofia. Experience in disaster and humanitarian relief operations. G Med Mil 162(1):133–136. http://www.difesa.it/GiornaleMedicina/rivista/Documents/Rivista_1_2012/133-136_MilitaryMedical_Academy
7. Kanev Kamen, Rostislav Kostadinov, Aleksander Parashkevov, Aleksandar Dimitrov. Civil-military cooperation in the medical universe. //17-th Congress of the Balkan Military Medical Committee, May 29–June 01, 2012, Belgrade, Abstract Book 2012, p 45

CIMIC in Medical Disaster Relief Operations

Rostislav Kostadinov

Abstract In the last three decades, the world is facing unprecedented increase in both disasters' frequency and severity. Enhancing the community resilience toward natural and man-made disasters is becoming a prerequisite for country, regional, and world development.

Provision of the efficient and rapid medical assistance to the affected population remains a big challenge and difficult task for disaster managers. Proper disaster medical support requires specific knowledge and skills and ability to adapt yourself to the particularities of the environment affected by the calamity.

The aim of this publication is to analyze why civil military medical cooperation and interaction is beneficial for the medical support to disaster relief operations.

Materials and Methods – Descriptive and comparative methods are applied for analyses of the characteristics of the disasters' impact on healthcare system and the requirements for the disaster medical support to the affected population that are matched with military medical support capabilities. Based on the obtained results in the conclusion by cluster analysis, some of the possible beneficial interactions between civil and military medicine in case of disaster relief operations are presented.

Keywords Disasters' impact • Disaster medical support • Military medicine • Civil military cooperation

Introduction

In the last three decades, the world is facing unprecedented increase in both disasters' frequency and severity. In accordance with the UNISDR, almost half of world population was affected by the disasters within last 12 years [1]. This trend is mainly due to climate change-related natural disasters, but communities nowadays have also to respond to the challenges related to man-made crises with almost disastrous

R. Kostadinov (✉)
Bulgarian Military Medical Academy, Sofia, Bulgaria
e-mail: rostikosti@abv.bg

O. Nikolov, S. Veeravalli (eds.), *Implications of Climate Change and Disasters on Military Activities*, NATO Science for Peace and Security Series C: Environmental Security, DOI 10.1007/978-94-024-1071-6_10

features as refugee and migrant flow; local military conflicts, caused by failing states; civil unrest; and international disputes. The 2011 Japan earthquake revealed that disasters are able to devastate every country, even those with highly effective and prosperous economy and well-organized and wealthy community. In the last years, the world is shocked by the increase in the terrorism activity and cruelty. Terrorist attacks are becoming almost daily news and the novel is not the act itself but the place, number of victims, and mode implicated. Gunshot and blast trauma injuries are common consequences of the terrorist activities. Recently, we have been witnessing terrorists attacks, where chemical weapons were implemented successfully.

Unfortunately, we could only predict that the recorded trend of increase in disaster frequency and severity will continue, because the origin of this increase could be found into the main processes characterizing our contemporary world – urbanization (increasing the population at risk, as well as risk level related to the biological hazards and transport accidents), industrialization (trigger for climate change and increasing the chemical and radiological threat and depletion of water and mineral resources), poverty, famine, and social inequality (that could lead to armed conflicts and terrorism).

Enhancing the community resilience toward natural and man-made disasters is becoming a prerequisite for country, regional, and world development.

Provision of the efficient and rapid medical assistance to the affected population remains a big challenge and difficult task for disaster managers. Increasing the disaster medical readiness level of the medical specialists, of rescue team members, and of the population as a whole could be defined as a milestone in community resilience. Proper disaster medical readiness requires specific knowledge and skills and ability to adapt yourself to the particularities of the environment affected by the calamity.

Military medicine was established in order to provide medical assistance to the injured warriors on and afterward close to the battlefield. There are several similarities between medical activities required by the particularities of the medical situation developed as consequence of the military operations and majority of disasters. Algorithms for medical management of the battlefield casualties on the field could be easily adapted to become basis for disaster medical education and training. Including military medical expertise into planning and management of the disaster relief operation will broaden their capabilities to respond adequately to the demands of gunshot and blast trauma mass casualty events, as well as events with CBRN environmental contamination, to note just a few.

The aim of this publication is to analyze why civil military medical cooperation and interaction is beneficial for the medical support to disaster relief operations.

Materials and Methods

Descriptive method is applied for analyses of the characteristics of the disasters' impact on healthcare system and the requirements for the disaster medical support to the affected population. By the means of comparative method, the disaster medical support requirements are matched with military medical support capabilities. Based on the obtained results in the conclusion by cluster analysis, some of the possible beneficial interactions between civil and military medicine in case of disaster relief operations are presented.

Results and Discussion

One event by definition is classified as disaster when there is disparity between required and available for its management resources that requires external assistance [2]. Appling this definition into medical domain, we could easily determine that the main characteristics of the disaster medical situation are related to the disaster damaging factors impact on the population health, medical infrastructure and healthcare system operability.

As the medicine is defined as science for diagnosis, prevention, and treatment of diseases, it is rational to expect that disaster medicine topic is the negative health consequence in case of disastrous event – injuries and diseases. One could argue that diagnosis, prevention, and treatment are the topics of the medicine as a whole; therefore, the procedures and algorithms of medicine should be sufficient for medical management of the disaster consequences. No one could deny such a statement – the main objective is to save the life and preserve the health and ability of the patient. The particular demands are due to the abovementioned into the definition feature – the lack or scarcity of resources required for situation management:

1. One of the main disaster features is their unpredictability – no one could assure the disaster medical support planners and managers where, when, what, and with what magnitude disaster will occur.
2. In case of disaster great, majority of the injuries are occurring almost simultaneously that results in number of victims highly exceeding the capabilities of healthcare system to respond.
3. Records present data that disasters' damaging factors predominantly are leading to severe injuries requiring immediate lifesaving medical aid or complex and sophisticated treatment.
4. Due to the abovementioned features, the disaster medical support in the beginning is always characterized by insufficient quantity of sanitary transport means, hospital room, and medical specialists – physicians, nurses, and technicians. The scarcity of technical and medical equipment required for proper diagnosis and treatment of the injuries is also notable.

5. Great majority of disasters are leading to damages of lines of communications – roads, air and seaports, phone, mobile, and internet connections are overwhelmed or with damaged elements that impede or limit their capacities to operate as expected.
6. The level of stress and panic among affected population, rescue teams, and society as a whole are other disasters' characteristic with impact on individual and community health.
7. Mass casualty events require specific medical organization and different approach toward patients, applying slightly different triage algorithm.
8. Very often disasters are setting austere and hostile to the rescue team members' health environment, where they are supposed to fulfill their duties; another challenge is whether to save first the casualty or to assure his/her safety, a question that is not taught in the university or has to be responded during daily practice. [2]

Based on the abovementioned disaster characteristics with impact on healthcare system, it is easy to highlight the main challenges to the civilian healthcare providers:

1. Difficulties to reach the casualties due to the damaged transport infrastructure
2. Setting the forward temporary medical station in accordance with the environmental hazards
3. Diverse casualty triage and management algorithms
4. Mass casualty situation management
5. Field working stations
6. Scarcity or lack of proper medical equipment for injury management
7. Limited to absent laboratory and imaginary support
8. Limited technical equipment
9. Constant health threats
10. High medical relief operation tempo
11. Long shifts
12. Insufficient medical personnel
13. Limited to absent possibilities for medical consultations and collaboration with other medical specialists
14. Great diversity of casualties – different types of injuries, age, sex, comorbidities, chronic or exacerbated chronic disease, etc.
15. A number of casualties with significant stress reactions
16. Working under pressure and in strict time limitations
17. Requirements of coordination and subordination with the rest of the rescue teams [3–7]

All these challenges and shortfalls have to be adequately responded in a timely manner, because in disaster-stricken area every time loss is equal to life loss.

Nowadays the medical science is making enormous leaps in its development by implementing almost on daily basis the highly sophisticated and elaborated technically medical equipment. This development is leading to unbelievable progress into treatment of the diseases and traumas. On the other hand, these technical means are limiting the freedom of the medical specialists to work in conditions, where they have to rely only on their knowledge and proper skills, as are the areas devastated by the natural or man-made calamities. This highly technical and equipment dependence of civilian medical specialist is one growing and still unsolved issue.

Military medicine is a branch of medical science, where medical specialists are taught and trained to work into the hostile environment of the battlefield. Due to the unpredictability of the military operation outcome, every single military medic is trained to manage mass casualty situation – not only how to manage the casualties flow but also how to set the triage, treatment, and evacuation areas. Because of the requirements for readiness for work on the field relaying on its own energy and water supplies, the military medical installations are equipped with simplified and limited technical equipment; therefore the medical specialist training is focused on how to save more lives by utilizing less technical and material resources. Moreover, the military sanitary transport is designed for reaching remote and difficult or inaccessible terrains. Furthermore, the military medical doctrine is clear regarding the need of teamwork in close command and control relation, as well as in coordination with the other military service branches.

Comparing the challenges disaster medical support is posing to civilian healthcare system and the system shortfalls with the capabilities and capacities of the military medicine, it is reasonable to note that the civil military cooperation in case of planning and execution of the medical assistance to the affected population could ameliorate the outcome of disaster medical relief operations. The reasons for this statement are that:

1. Military medical specialists with their training to set field hospitals could benefit the forward temporary medical station establishment, as well as the organization of the work flow within it.
2. Military medical algorithms could be easily adapted to the requirements of the particular medical situation, caused by the disasters' damaging factors and impact. Military sanitary transport capabilities will limit the time required for medical assistance provision.
3. Military medical specialist inclusion in disaster medical teams could provide the team with practical solutions for problems related to the technical and medical equipment absence.
4. Specialized training for managing the battlefield stress reactions and posttraumatic stress disorders, as well as how to deal with the first signs of the burn-out syndrome, medical specialists are receiving is of utmost importance into the first stage of population stress and panic management [8, 9].

Conclusion

All these analyses lead to the conclusion that civil military cooperation and interaction is not only desirable but mandatory for better disaster medical support operation management and outcome. In order to smooth the process of interaction, it is recommendable that joint civil military medical disaster relief theoretical seminars and computer-based or life exercises are organized and executed. What could further benefit the result of such a cooperation is the level of these theoretical and practical events to be decreased from the national to municipal and provincial level, thus forming a real standing medical teams that will know each other and could rely on others' knowledge and capabilities.

References

1. https://www.unisdr.org/we/inform/disaster-statistics
2. Kostadinov, R. 600 simple steps for disaster medicine exam success. VAP Publishing House, Plovdiv, 2012 r. 176 p. ISBN: 978-954-8326-56-8
3. Ford ES, Mokdad AH, Link MW et al (2006) Chronic disease in health emergencies: in the eye of the hurricane. Prev Chronic Dis 3(2):1–7
4. Nickerson JW, Hatcher-Roberts J, Adams O, Attaran A, Tugwell P (2015) Assessments of health services availability in humanitarian emergencies: a review of assessments in Haiti and Sudan using a health systems approach. Confl Heal 8:920
5. Burkle FM, Redmond AD, McArdle DF (2011) An authority for crisis coordination and accountability. Lancet 379(9833):2223–2225
6. VanRooyen M, Leaning J (2005) After the tsunami – facing the public health challenges. N Engl J Med 352(5):435–438
7. Lind K, Gerdin M, Wladis A, Westman L, von Schreeb J (2012) Time for order in chaos! a health system framework for foreign medical teams in earthquakes. Prehosp Disaster Med 27(1):90–93
8. Kostadinov R, Dimitrov A (2012) Military Medical Academy, Sofia experience in disaster and humanitarian relief operations. G Med Mil 162(1):133–136. http://www.difesa.it/GiornaleMedicina/rivista/Documents/Rivista_1_2012/133_136_Military_Medical_Academy.pdf
9. Joob B, Wiwanitkit V. (2014) Acute natural disaster relief by role of military medicine. J Acute Dis:165–166

Condition and Impacts of Climate Change

Craig Hanrahan

Abstract The State of Tennessee takes an all-hazards approach when responding to, preparing for, mitigating against, and recovering from disasters. This approach is articulated in the Tennessee Emergency Management Plan (TEMP) and its annexes, the Tennessee Catastrophic Annex (TNCAT) and the Tennessee Hazard Mitigation Plan (HZMIT). Because Tennessee uses an "all-hazards" approach, these plans are used to address the amplified impacts that result from climate change.

Keywords National Flood Insurance Program • Mitigation plan • Climate change

All-Hazards Approach and View on Climate Change

The State of Tennessee takes an all-hazards approach when responding to, preparing for, mitigating against, and recovering from disasters. This approach is articulated in the Tennessee Emergency Management Plan (TEMP) and its annexes, the Tennessee Catastrophic Annex (TNCAT) and the Tennessee Hazard Mitigation Plan (HZMIT). Because Tennessee uses an "all-hazards" approach, these plans are used to address the amplified impacts that result from climate change.

Federal guidelines required that the 2013 Tennessee Hazard Mitigation Plan update to address "climate change." Below is the reference to climate change in the State HZMIT Plan. It is currently the only reference. For 2018, there will be a more in-depth examination.

One additional item of discussion which took place was climate change. In recent years some states have started to address climate change as a hazard of prime concern. After various decisions, committee representatives decided not to profile climate change in this year's State Hazard Mitigation Plan Update for two main reasons. First, some committee members deemed climate change as more of a "condition" than a "hazard." For example, it is a condition because it amplifies the impacts and occurrences of already existing natural hazards of prime concern.

C. Hanrahan (✉)
Bureau of Response, Tennessee Emergency Management Agency, Nashville, TN, USA
e-mail: chanrahan@tnema.org

O. Nikolov, S. Veeravalli (eds.), *Implications of Climate Change and Disasters on Military Activities*, NATO Science for Peace and Security Series C: Environmental Security, DOI 10.1007/978-94-024-1071-6_11

Second, the committee decided that the state should wait for formal federal guidance on how to address climate change in mitigation plans before fully incorporating it. Because of continuing climate change debates and unaddressed questions *(such as what types of climate change mitigation actions qualify for FEMA mitigation grants and how does a state conduct a state-specific cost-benefit analysis (CBA) to justify climate change mitigation actions?)*, the state will wait for further federal guidance.

As articulated in the statement above, climate change is currently viewed as a condition that amplifies a given threat/hazard.

Conditions of Climate Change

This section of the presentation attempts to show the open-sourced global and regional data that is available to highlight the conditions of climate change. According to the National Oceanic and Atmospheric Administration (NOAA), significant climate anomalies have been identified globally including the contiguous United States (severe flooding across Central and Eastern Texas) and North America (warmest May since records began in 1910). This information can be found by visiting https://www.ncdc.noaa.gov/sotc/global/201602.

Flooding has always been a hazard of prime concern. However, in reviewing the percent change of very heavy rain events in the United States, it is clear that there has been a very significant increase from 1958 to 2012, 27% in the region that the State of Tennessee is situated. The actual rain event may not be as frequent, but the accumulations are higher resulting in flood events (http://nca2014.globalchange.gov/highlights/overview/overview).

In addition to increased flooding events, another NOAA dataset graphic (Land and Ocean Temperature Percentiles, May 2016) reveals that May 2016 had a global land surface temperature of 1.57 F above the twentieth-century average of 58.6 F (https://www.ncdc.noaa.gov/sotc/global/201605). Further trends show that the United States is only going to continue getting warmer.

The impacts resulting from increased heavier rainfalls followed by periods of drought are the increased severe flooding events, agricultural disasters, and forest fires in the State of Tennessee.

Climate Impacts in Tennessee

In May of 2010, Tennessee experienced an extended flash flood that dropped up to 20 in. of rain in areas ranging from Memphis to Nashville in a 48 h period. This caused extreme flooding in rivers, creeks, storm-water drainages, and other low-lying areas and resulted in a federally declared disaster. The capital city of Nashville and other rural and metropolitan areas experienced major damages.

In May of 2011, almost 1 year to the day of the 2010 flood, a large storm system stretching from southern Texas to the Great Lakes dump very heavy amounts of rain causing a slow rise of the Mississippi River. Although the river rose slowly and gave communities time to prepare, flooding was tremendous and resulted in another federally declared disaster.

In April of 2016, an unseasonably dry spring results in a series of forest fires in the eastern region of Tennessee. There were 18 wildfires that impacted 11 counties and a national park. These fires stretched local and state resources and federal assets were utilized.

Overview of Floodplain Management in the United States

Floodplain management is the operation of a community program of preventive and corrective measures to reduce the risk of flooding (http://www.fema.gov/floodplain-management). In the United States, floodplain management is supported by all levels of government, but is primarily the responsibility of the local government. There are three key components underlying the US approach to floodplain management: data, policy and regulation, and incentives.

To gather data, the federal government has conducted flood studies of populated areas in the United States. The data from the studies is used to develop flood risk maps, which are publically available both online and through the local government (https://msc.fema.gov/portal). To date, almost 100,000 of these maps have been developed. These maps and data are also the basis for insurance companies to provide flood insurance through the National Flood Insurance Program (NFIP).

Policy and regulations are based on the data gathered from flood studies and data from studies of specific areas of focus, such as climate change and sea level rise. At the federal level, all federal agencies are tasked by the President to take action to reduce the risk of flood loss (http://www.fema.gov/executive-order-11988-floodplain-management-0). Any federal agency that provides funding, permitting, or other support to any construction or development must ensure that flood risk is minimized for that project. Many states and communities also promulgate additional flood risk reduction policies and regulations that are more specific to address the risks of the region.

Incentives such as insurance rate discounts are available through the National Flood Insurance Program (NFIP) for communities and individuals meeting a high standard of flood plain management. This program is known as the Community Rating System (CRS). More than 1200 communities from all 50 states participate in the CRS. In the past 5 years, all 50 US states have experienced floods or flash floods, and there are currently more than 5.3 million flood insurance policies in force across more than 22,000 communities in the United States. Effective floodplain management will continue to grow in importance as climate change, increased development, and population growth increase flood risk worldwide.

Partnerships

In addition to floodplain management policies, focused and deliberate partnerships are critical to help lessen the impact of amplified threats/hazards due to climate change.

From 2010 to 2012, the State of Tennessee weathered eight major disasters resulting in presidential declarations, including record-breaking floods in several Tennessee watersheds. Stakeholders needed innovative solutions to help address burgeoning disaster risk in the state. In September 2013, a small group of those stakeholders hosted the official kickoff of a collaboration designed to find and foster those innovative solutions: *Tennessee Silver Jackets*. A year later, 18 different agencies signed the official charter of the *Tennessee Silver Jackets*, including the mayors of the five most populous cities and metropolitan areas in the state. In less than 3 years, a group of a few concerned stakeholders has grown into a recognized and productive team.

Following a large-scale disaster (tornadoes, floods, earthquakes, etc.), there is a need to assess buildings and structures quickly to ensure safe response and reentry. Individuals will want to return to their homes and businesses. The Tennessee Structural Assessment and Visual Evaluation (TNSAVE) Coalition was created to help develop and maintain a post-disaster building inspector program. This group of professional organizations consists of professional engineers, architects, and building inspectors.

Disaster Risk Management: Using a Resilience Systems Approach to Plan for Multiple Stressors

Janani Vivekananda

Abstract Though the interconnections between disasters and security are becoming better understood, governments and militaries remain poorly equipped to assess and manage the risks that climate change and disasters pose to security and development. The identification of these risks constitutes only the first step in the process of implementing a risk management approach that effectively reduces vulnerabilities and builds the resilience of state institutions and societies. The next step is deeper analysis of likelihood, severity and variability for each of these risks along a range of climate change scenarios while also situating them within specific geographic, socioeconomic and political contexts. For this, it is necessary to look not just at climate hazards but to the systems within which they play out. A resilience systems analysis provides one way of doing this at a macro level. At the same time, a conflict-sensitive risk analysis enables an understanding of appropriate responses at a local level.

Keywords Disaster risk reduction • Risk management • Resilience • Resilient systems • Risk • Conflict • Peace • Climate change

Disaster Risk Management: Using a Resilience Systems Approach to Plan for Multiple Stressors

Presentation Overview and Objective

This presentation sets out two practical approaches to assess combined climate change, disaster and conflict risks and shape appropriate responses at different scales. It first outlines the compound nature of disaster risk and security. It then offers a brief overview of a resilience systems approach and a conflict-sensitive approach and how they help look at climate and conflict interactions at different levels.

J. Vivekananda (✉)
Adelphi, Berlin, Germany
e-mail: vivekananda@adelphi.de

O. Nikolov, S. Veeravalli (eds.), *Implications of Climate Change and Disasters on Military Activities*, NATO Science for Peace and Security Series C: Environmental Security, DOI 10.1007/978-94-024-1071-6_12

Presentation Key Takeaways

Much is now known a great deal about different risks in vulnerable countries. There are numerous multiple-criteria risk analysis tools, showing us where and when conflict is likely and which areas are exposed to natural disasters and modelling how economic shocks and pandemics might spread or how climate change will affect different communities and regions.

However, a recent analysis of 66 conflict and environmental risk assessment tools showed only three looked at climate change, disaster and conflict risks together and none did so in a meaningful way [1]. Without tools which enable analysis of interacting risk factors, we don't yet share a vision of what to do about those risks and how to boost the resilience of individuals, households, particular communities and states to the risks they face.

A resilience systems analysis is attractive because it provides key actors in the field with:

- A shared view of the risk landscape that people face
- An understanding of the broader system that people need for their all-round well-being
- An analysis of how the risk landscape affects the key components of the well-being system and which components are resilient, which are not and why
- A shared understanding of power dynamics and how the use or misuse of power helps or hinders people's access to the assets they need to cope with shocks
- A shared vision of what needs to be done to boost resilience in the system and how to integrate these aspects into policies, strategies and development efforts at every layer of society

Resilience systems analysis uses a systems approach. This is because the impact of the risk is dependent on how society's systems are set up to respond to shocks and change. The analysis process:

- Starts with an understanding of the risk landscape in a particular context
- Looks at how those risks will affect society's systems
- Gathers information about how those systems are set up to cope with those risks and whether this:
  - Makes them resilient
  - Determines what needs to be done to boost resilience and to help the different parts of the system to:

    Either absorb those shocks or adapt so that they are less exposed to those shocks or transform so that:

    The shock will no longer affect them.

    The result is a resilient system, which will then change the overall context and risk landscape.

Boosting resilience involves:

- Actively understanding the risk landscape and how it impacts on systems – how society functions in:
  - Each context
- Determining at which layer of society those risks are best managed

The purpose of resilience systems analysis is to inform policies, strategies and programmes, rather than to support detailed project design. Given the breadth of the risk landscape examined, a resilience systems analysis can only develop an aggregate vision of strengths and weaknesses across the system. As such, at the operational level, this approach needs to be supplemented with a community-level conflict-sensitive context analysis.

Conflict sensitivity entails awareness, management and mitigation of the unintended consequences of an action on the context that may create or exacerbate conflict. In order to avoid these risks and indeed to positively promote peace opportunities in a given context, actors must have a thorough understanding of conflict drivers, the power holders and dynamics and the political economy of a given context and how they interact at a local level. The importance of understanding this interaction at a local level is underpinned by the numerous recent environmentally induced humanitarian emergencies which interact with ongoing or latent conflict, for example, Syria, South Sudan, Mali and Kenya.

In addition, central to a conflict-sensitive approach is the need for a better understanding of root causes and broad and inclusive engagement with those who understand and are affected by risks at the centre of the response. This is fundamental if we are truly to shift towards a focus on prevention and management rather than almost exclusively on response.

Creating opportunities for dialogue and community engagement can open channels for complaints and grievances to be voiced – enabling responders to troubleshoot potentially misdirected intervention efforts. Importantly, dialogue and community engagement are key to building trust among communities and between communities and the government. Trust may not be a traditional disaster management outcome, but is core to sustainable impacts, especially in conflict-affected states.

Such an approach will facilitate disaster management efforts to address the root causes of vulnerability to disaster impacts. And addressing these root causes is perhaps the ultimate 'no regrets' phase zero intervention – for example, addressing issues such as lack of livelihood diversification, political marginalisation, weak and inflexible institutions and inequitable policy processes. These can help communities and governance providers plan for uncertainty and peacefully manage a range of possible futures which climate change presents. Within such a process though, it is important to remember whose risk you are talking about. A risk for one group of stakeholders may not be for another. For example, elites controlling a scarce resource such as water may indeed profit from scarcity of a natural resource as they are able to drive up prices.

Conclusion

The risks posed by climate change are complex and require an understanding of potential second- and third-order consequences. To address these challenges, we need to ensure that any intervention in a fragile state takes account of conflict risks and climate risks and accept that unless we address medium- and long-term climate risks in short-term humanitarian and disaster management approaches, these responses will not be sustainable in a changing climate and thus not promote lasting peace. In fact, they can do more harm than doing nothing.

Focusing on the climate hazard is one way of broaching the issue, but a better one is to address the factors which render people vulnerable to these hazards and build resilience to prevent or adapt to these impacts. The complexity and simultaneous local and transboundary nature of the challenge of disasters and security require action on a dual track.

At the transnational level, this means moving towards global-/transboundary-level resilience systems risk assessments. Governments carry out many different risk assessments, often high quality, but these are often disjointed and not coordinated within government departments, let alone across different governments. A whole-systems approach to measuring compound climate-fragility risks would require integration of four forecasting segments: greenhouse gas emissions; the climate systems' response; the knock-on consequences on society, economy and politics – including the conflict and fragility risks that arise from them; and the consequences of responses designed to address climate risks. It will need to engage governments, militaries and communities.

But localised understandings of and responses to risk are also vital. Assessments which look at community resilience to climate/disaster-security risks will help pixilate the highly nuanced risk picture at the national and international level and enable more targeted and effective responses.

Reference

1. Peters K, Vivekananda J Topic guide: conflict, climate & environment, November 2014, evidence on demand

Literature

2. OECD, 2014, Guidelines for resilience systems analysis. OECD publishing. https://www.oecd.org/dac/Resilience%20Systems%20Analysis%20FINAL.pdf
3. Africa Peace Forum, CECORE, CHA, FEWER, International Alert, and Saferworld (2004) 'Conflict- sensitive approaches to development, humanitarian assistance and peacebuilding' – a resource pack, January 2004. http://www.internationalalert.org/publications/140.php

Disaster Risk Management: Assessing and Leveraging Existing Capacities

Carl Bruch

Abstract This chapter highlights existing institutions and capacities that can be deployed for disaster risk management. It starts with a discussion of the process for assessing capacities and some of the key international, regional, and national actors in the military and civilian sectors with technical expertise, capacity, and authority to address climate change-related natural disasters, including those that work on planning, preparedness, response, and recovery. It then considers tools for building adaptive capacity, including a course on "Climate Change, Disasters, and Security" and a suite of tools on environmental peacebuilding.

Keywords Capacity assessment • Institutions • Training • Environmental peacebuilding • Disaster risk management

Presentation Overview and Objectives

This chapter highlights existing institutions and capacities that can be deployed for disaster risk management. It starts with a discussion of the process for assessing capacities and some of the key international, regional, and national actors in the military and civilian sectors with technical expertise, capacity, and authority to address climate change-related natural disasters, including those that work on planning, preparedness, response, and recovery. It then considers tools for building adaptive capacity, including a course on "Climate Change, Disasters, and Security" and a suite of tools on environmental peacebuilding.

C. Bruch (✉)
Director of International Programs, Environmental Law Institute, Washington, DC, USA
e-mail: bruch@eli.org

O. Nikolov, S. Veeravalli (eds.), *Implications of Climate Change and Disasters on Military Activities*, NATO Science for Peace and Security Series C: Environmental Security, DOI 10.1007/978-94-024-1071-6_13

Key Takeaways

There are two basic approaches to assessing capacity to address climate change-related natural disasters: supply and demand. In considering the supply side, it is important to examine which are the relevant institutions and what capacities they have. In considering the demand side, in contrast, the focus is on the needs (rather than what exists), and the questions focus on which capacities are necessary and which institutions might have those capacities. In practice, assessment of capacities usually addresses both the supply and demand side, comparing the existing capacities with those that are needed.

The assessment process should consider a range of capacities. These include capacities across the disaster risk management cycle, addressing prevention (including planning), mitigation, preparedness, response and relief, and recovery. It should also include science/data, policy, and operational expertise in both the military and civilian sectors.

While there are a range of possible assessment methodologies, they tend to emphasize three key considerations. First, the process should be participatory, engaging a broad cross section of military, governmental, academic, and civil society institutions. Second, it is important to think creatively about the various institutions and relevant needed capacities. Finally, in assessing capacity, consider existing linkages for dialogue, coordination, and cooperation.

The result of an assessment will likely identify numerous and varied relevant institutions, including:

- International actors, such as various UN agencies, other international organizations (such as the World Bank), and foreign militaries and aid agencies
- Regional actors, such as NATO and the European Union (including the Emergency Response Coordination Centre (ERCC) and Common Emergency Communication and Information System (CECIS))
- National government actors, including both military (army, national guard, air force, avy) and civilian (disaster management agencies, planning agencies, environment agencies, etc.) authorities
- Other stakeholders, including civil society, academia, the private sector (including insurance companies), and local authorities.

Growing disaster risks and limited capacity mean that successful disaster risk management (DRM) requires a whole-of-government approach. In such approaches, multiple military and civilian government agencies cooperate closely on DRM activities. In many countries, militaries have significant (and sometimes greater) disaster preparedness and response capacity than civilian agencies. As a result, disaster response can be greatly increased by deploying military assets, including equipment and personnel. To make this work, though, it is necessary to have the requisite policies and operating protocols in place to ensure effective civ-mil coordination.

The Oslo Guidelines on the Use of Military and Civil Defense Assets in Disaster Relief provide a basic framework for formalizing and improving the effectiveness of military assets in disaster response [2]. The core principles of the guidelines are humanity, neutrality, impartiality, and operational independence, with an emphasis on deference to civilian authority.

The US Army Corps of Engineers, the Africa Command of the US Army (AFRICOM), the Joint UNEP/OCHA Environment Unit, and the Environmental Law Institute have developed a training course on "Climate Change, Disasters, and Security." The course seeks to build capacity in African countries and partner organizations to develop whole-of-government approaches for natural disaster prevention and mitigation, preparedness, response, and recovery, focusing on disasters that are a consequence of or that are exacerbated by climate change. The course targets African civil and military authorities, partner organizations, local and national emergency management agencies, and civil society. Through the course, participants learn about linkages between disasters and security and gain practical tools to improve prevention and mitigation, preparedness, response, and recovery to climate-related natural disasters. Elements of the course were pilot tested in a November 2015 workshop in Dakar, Senegal, and there is growing interest in other regions in adapting the course to their particular context.

Environmental peacebuilding also presents tools to assess and leverage adaptive capacities. "Environmental peacebuilding" is the process of governing and managing natural resources and the environment to support durable peace [1]. It comprises actions across the conflict life cycle, including efforts to prevent, mitigate, resolve, and recover from violent conflict; and it can involve efforts related to renewable natural resources (such as land, water, and fisheries), nonrenewable natural resources (such as minerals, oil, and gas), and ecosystems (including their services).

A wealth of environmental peacebuilding tools may be utilized to build national and local capacities to manage the effects of climate change, manage disaster risks, and promote security. These include the Environmental Peacebuilding Knowledge Platform (www.environmentalpeacebuilding.org), an Environmental Peacebuilding Community of Practice (with more than 2900 people in more than 90 countries), an *Environmental Peacebuilding Update* (that is produced every 2 weeks), an Environmental Peacebuilding Academy (for teachers and trainers), an annual lecture and award on environmental peacebuilding, operational guidance, and a massive open online course (MOOC) that is under development.

Presenter Background

Carl Bruch is the Director of International Programs at the Environmental Law Institute. He has helped countries and organizations throughout Africa, America, Asia, and Europe develop and strengthen their environmental laws, improve institutions, and build capacity. His work focuses on environmental peacebuilding (especially after conflict), environmental governance, adaptation, and environmental emergencies.

Suggested Key Readings

See References.

Conclusion

Effectively dealing with the security risks of climate change-related disasters requires a whole-of-government approach and indeed a whole-of-society approach. Key to developing an integrated and coordinated approach is assessing and leveraging military and civilian capacities. Assessment relies on both supply-oriented and demand-oriented analyses of existing institutions at the international, regional, national, and other levels. Leveraging capacities may be accomplished through a training course on "Climate Change, Disasters, and Security" and through various environmental peacebuilding tools.

References

1. Bruch C, Jensen D, Kron A (2016) Environmental peacebuilding. Understand to prevent: guidance on the military contribution to the prevention of violent conflict. Multinational Capability Development Campaign, pp 84–88
2. UNOCHA (UN Office for the Coordination of Humanitarian Affairs) (2007) Oslo guidelines on the use of military and civil defense assets in disaster relief. Rev. 1.1. Geneva

Geoinformatics for Disaster Risk Assessment, Monitoring, and Management

Andmorgan R. Fisher

Abstract Over the past years, there has been a shift in focus from "disaster recovery and response" to "risk management and mitigation," and the use of geoinformatics has become an integrated, well-developed, and successful tool in disaster risk management. Geoinformatics technologies include communication and information technologies coupled with geographic information systems, or GIS, GPS, and remote sensing. There are many technologies that can be used to facilitate mitigation, preparedness, response, and recovery when dealing with natural disasters due to climate issues. However, the most effective programs will require coordination, cooperation, and communication. Technologies are developing rapidly while costs are falling, and disaster relief efforts stand to benefit.

Keywords Remote sensing • Geoinformatics • Disaster risk management

Presentation Overview

Major disasters, such as the 2004 Indian Ocean tsunami and the 2011 Pacific Ocean tsunami, have highlighted challenges with respect to preparedness that many areas face when dealing with events of large magnitude [1]. Climate change is exacerbating these types of challenges by increasing the frequency and variability of extreme weather events. Geoinformatics technology can potentially assist in disaster risk assessment, monitoring, and management. Geoinformatics technologies include communication and information technologies coupled with geographic information systems, or GIS, GPS, and remote sensing.

Disaster risk management is the systematic process of using administrative decisions, organization, operational skills, and capacities to implement policies, strategies, and coping capacities of communities to reduce the impacts of natural hazards and related environmental disasters [2]. Disaster management includes all measures

A.R. Fisher (✉)
Geospatial Research Laboratory, Engineer Research and Development Center,
US Army Corps of Engineers, Washington, DC, USA
e-mail: andmorgan.r.fisher@usace.army.mil

O. Nikolov, S. Veeravalli (eds.), *Implications of Climate Change and Disasters on Military Activities*, NATO Science for Peace and Security Series C: Environmental Security, DOI 10.1007/978-94-024-1071-6_14

taken to prevent or limit the adverse effects of disasters in the pre-disaster, or preparedness and mitigation, phases and in the post-disaster, or response and recovery, phases. Over the past years, there has been a shift in focus from "disaster recovery and response" to "risk management and mitigation," and the use of geoinformatics has become an integrated, well-developed, and successful tool in disaster risk management.

At the mitigation stage, geoinformatics can be of help with respect to risk modeling, vulnerability analysis, land and water management planning, hazard mapping, land use mapping, and digital elevation models. Drought, for example, and the famine it can cause, is a major disaster in the developing world, particularly in Africa. Unlike many natural disasters, famine sets in slowly and can often be predicted months in advance. Long-term climate forecasts, derived from satellite observations, can help build various scenarios before or during the early crop-growing season.

For example, the Famine Early Warning Systems Network, or FEWS NET, funded by the USAID monitors food security via satellites. It uses vegetation indices calculated from sensors including advanced very-high-resolution radiometer (AVHRR), moderate-resolution imaging spectroradiometer (MODIS), and Satellite Pour I'Observation de la Terre (SPOT) to monitor vegetation vigor and density, thus enabling problems to be identified as they develop. Rainfall is estimated using Meteosat infrared data and combined with rain gauge reports and microwave satellite observations. The satellite data is combined with regional analyses of prices, grain stores, political conditions, and agricultural inputs to provide effective early warning of potential food shortages. Other initiatives using satellite data for drought monitoring and famine prediction include the Agricultural, Hydrological, and Meteorological Programme (AGRHYMET) in West Africa, the IGAD Climate Prediction and Applications Centre (ICPAC) run by East Africa's Intergovernmental Authority on Development, and the Southern African Development Community's Regional Remote Sensing Unit.

An example of how geoinformatics can be helpful in the preparedness stage is flood monitoring. At later stages of the disaster cycle, such data can also be used to inform response and recovery efforts. The Global Precipitation Measurement (GPM) mission is cosponsored by NASA and the Japan Space Exploration Agency and consists of an international network of satellites that provides next-generation global observations of precipitation from space. GPM data combined with land surface data provides improved weather forecasting; forecasting of floods, hurricanes, and droughts; integrated hydrologic models of watersheds; and climate models.

At the response stage, fire mapping provides an example of how thermal data from MODIS can feed into a fire sensorWeb, which can in turn be used to identify hotspots and potentially trigger other collection requests. MODIS produces global fire maps that show active fires over the past 10 days. Figure 1 depicts the MODIS fire map from July 5, 2016. The active fire mapping system is used by a wide array of fire monitoring programs.

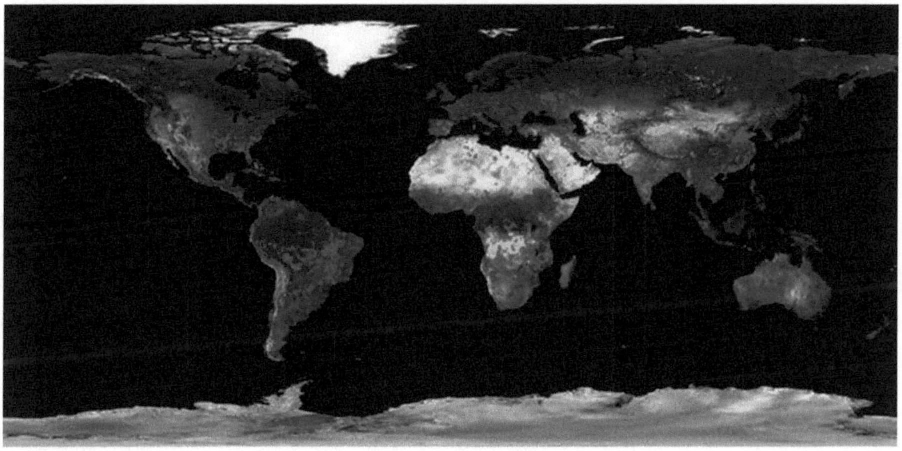

Fig. 1 MODIS fire data from July 5, 2016

With respect to recovery, geoinformatics can aid in damage assessment, spatial planning, informing drought mitigation, identifying sites for rehabilitation, and suggesting management practices. In October 2015, the Buckeye platform, which is owned by the US Army Geospatial Center was deployed for its first use for Humanitarian Assistance and Disaster Relief (HA/DR). Buckeye is a fixed-wing collection platform that provides unclassified 10 cm imagery. The imagery provided by Buckeye provided valuable and timely information to the Armed Forces of the Philippines.

More cost-effective options for rapid imagery collection include the use of small unmanned aerial systems, such as the Sensefly ebee fixed wing mapping drone. The Geospatial Research Laboratory currently uses such a platform for tactical mapping and has just started to explore using this technology for HA/DR applications. Like Buckeye, it produces centimeter scale data which is fairly simple to process and can be disseminated on the day of collection. Limitations of the system may include, however, use during extreme weather, flight altitude, and distance which the UAS can fly from the base station.

Key Takeaways

There are many technologies that can be used to facilitate mitigation, preparedness, response, and recovery when dealing with natural disasters due to climate issues. Ultimately however, the most effective programs will require coordination, cooperation, and communication. Technologies are developing rapidly while costs are falling, and disaster relief efforts stand to benefit.

Conclusion

The magnitude and frequency of climate-related natural disasters is increasing. Effective disaster management and response requires rapid utilization of information and data from many sources. Low-cost and free geoinformatics technologies can be used to facilitate mitigation, preparedness, response, and recovery. However, it is important to account for local solutions in many situations based on best-fit technologies and options for self-mobilization that may be crucial in many developing countries. The current rapid pace of technology development coupled with falling costs stands to greatly aid disaster relief efforts. The most effective programs, however, depend on coordination, cooperation, and communication.

References

1. Leidig M, Teeuw R, Gibson A (2013) The effectiveness of low-cost geoinformatics for disaster risk applications in coastal regions IEEE, pp 3698–3701
2. Van Westen CJ, Bakker WH, Andrejchenko V, Olyazadeh RW (2014). Risk changes: a spatial decision support system for analysing changing hydrometeorological risk. In: Proceedings of the international conference on analysis and management of changing risks for natural hazards, pp EO5–114

How Are Climate Change and Human Security Interrelated

Janani Vivekananda

Abstract The consequences of climate change are already upon us and are projected to worsen. Impacts are already being felt around the world. These impacts will be most harmful to security and stability in already fragile contexts where vulnerable communities are already facing the risk of conflict and insecurity. This presentation sets out seven compound risk factors through which climate change and human security are interrelated. This is intended to help the understanding of the current and future implications of climate change on human security in fragile states so that we can better address some of these risks through preventative interventions and resilience building where possible.

Keywords Climate change • Conflict • Human security • Risk • Resilience • Water • Food • Livelihoods • Adaptation • Mitigation • Conflict sensitivity

Presentation Overview and Objective

Climate change multiplies threats to the security environment. Addressing climate risks can increase peace and stability [1]. Yet at the global level there are a number of significant and immediate and seemingly more urgent challenges to face, most notably Russia's engagement in Ukraine, the turmoil in Syria and the broader Mediterranean region and terrorist attacks within NATO member states. However, it is important to note that climate change is not necessarily distinct from these other security risks, but rather, an important factor placing strains on the broader security environment in which these risks play out. Climate change will multiply stresses on factors related to peace and stability, such as water, energy, food and migration – all of which are key elements of instability within NATO states and in areas of strategic importance to NATO members, such as the Sahel and North Africa. NATO and other security providers need to be able to address more than one risk at a time. To do so requires a better understanding of the nature of these risks in order to inform

J. Vivekananda (✉)
Adelphi, Berlin, Germany
e-mail: vivekananda@adelphi.de

O. Nikolov, S. Veeravalli (eds.), *Implications of Climate Change and Disasters on Military Activities*, NATO Science for Peace and Security Series C: Environmental Security, DOI 10.1007/978-94-024-1071-6_15

preparedness efforts. This presentation sets out seven ways in which climate change interacts with pre-existing social, economic and political risks and identifies some potential entry points for preventative actions.

Presentation Key Takeaways

Risks are compound and can only be understood by looking at their interactions. The most meaningful way to understand climate change with regard to security is as a 'threat multiplier' which will interact with existing social, political and economic stresses – things like unemployment, inequality or lack of basic service provision, shifting the 'tipping point' at which conflict ignites. Compound risks are complex and interrelated with interaction and feedback loops between the risks. A recent report commissioned by the G7, A New Climate for Peace [2], sets out seven compound climate risks which affect fragile states. These risks are set out below.

1. Local resource competition: As the pressure on natural resources such as land and water increases – due to climate events, population growth and increased consumption as the global population develops – competition can lead to instability and even violent conflict in the absence of effective dispute resolution.
2. Livelihood insecurity and migration: Climate change will increase the insecurity of people such as farmers or herders who depend on national resources such as farm land or grazing land for their livelihoods. This could push them to migrate or turn to illegal sources of income – e.g. coffee farmers in Honduras are increasingly falling into organised crime. While migration can be an effective way to cope with climate stress, if migration and resettlement are poorly managed, it can lead to local and regional instability.
3. Volatile food prices and provision: Climate change is already affecting food production in many regions, increasing prices and market volatility and heightening the risk of protests, rioting and civil conflict. We are already seeing decreased crop yields and disrupted food production in many areas. Combined with other global stresses such as population growth and changing energy demands, food prices are only going to become more volatile. In 2007–2009, food price volatility leads to riots in over 40 countries around the world and can arguably be said to have been a trigger for the Arab Spring domino effect starting in Tunisia.
4. Extreme weather and disaster: Will exacerbate fragility challenges and can increase people's vulnerability and grievances – especially in conflict-affected contexts, where a state is unable to adequately respond to the risk. Disasters endanger and destroy people's livelihoods, assets, health and social networks. Most affected are often the poorest whose livelihoods and assets are the most fragile. The relationship between disasters and insecurity is often mutually reinforcing. Disasters put extra pressure on already stretched governance systems, decrease economic opportunities, reduce resources and displace people. A lack of safety nets, preparedness, insurance mechanisms and other coping methods

can fuel grievances – especially if responses are inadequate or inequitably distributed.
5. Transboundary water management: Transboundary waters are often a source of tension. As demand grows (through, inter alia, population growth and increased consumption) and climate change and disasters affect the availability and quality, competition over water use will likely increase the pressure on existing governance structures.
6. Sea-level rise/coastal degradation: Rising sea levels will threaten the viability of low-lying areas – even before they are submerged. This will lead to social disruption, displacement and migration, while disagreement over maritime boundaries and ocean resources may also increase.
7. Unintended impacts of climate policies: The final risk factor is not a result of climate change but a result of poorly planned responses. As climate adaptation and mitigation policies are more broadly implemented, the risks of unintended negative effects – especially in fragile contexts – will also increase. For example, the European Union's policy to move towards 20% renewable energy by 2020 (through boosting biofuels) contributed to food riots in over 40 countries around the world due to a concomitant spike in food prices.

Across all these risks, the most pertinent takeaway is an awareness of the interlinkages and the knock-on consequences of one risk on others. No risk is experienced in a vacuum; they will be shaped by other contextual factors such as poverty, weak governance or a history of conflict. For example, to take the case of Syria, the climate change-related drought from 2006 to 2011 overwhelmed the coping capacity of rural farmers, pushing vast swathes of rural-urban migration into already stressed urban centres such as Dar'aa. However, it was against the backdrop of a crumbling social contract and historic political grievances that tensions ignited. A risk assessment which only looked at climate risk or disaster risk would not have picked up the risk of conflict. And a conflict risk assessment would not have picked up the contributory role of the drought.

Conclusion

Climate change is a 'risk multiplier' which interacts with pre-existing social, economic and political risks, making peace harder to achieve. The flipside to this argument provides signposts for effective responses: through addressing compound risks – for example a post-conflict urban reconstruction initiatives which takes account of future changing water availability – it is possible to reduce risk by jointly building resilience to conflict, poverty and climate change [3]. There are indeed multiple and interconnected layers of uncertainty involved in planning for climate change in fragile states. Fundamentals include how much average temps will increase, what the impact of rapid climate change will be *and* how effective countries will be in agreeing to and implementing adaptation and mitigation plans.

Uncertainty cannot be a barrier to action. Uncertainty doesn't mean we know nothing, just that we don't know exactly what the future may hold. But this is the case in a range of public policy arenas, for example, military procurement, financial systems regulation and setting interest rates. The range of uncertainty in climate change is generally smaller than that common in long-term security analysis. As such, the strongest course of action is to strengthen governance capacity and institutions to be able to manage a range of uncertain possible futures.

References

1. Peters K, Vivekananda J Topic guide: conflict, climate & environment, November 2014, evidence on demand
2. Rüttinger L, Stang G, Smith D, Tänzler D, Vivekananda J (2015) A new climate for peace – taking action on climate and fragility risks
3. Vivekananda J, Schilling J, Smith D (2014) Climate resilience in fragile and conflict-affected societies: concepts and approaches. Dev Pract 24(4):487–501

Climate Change Implications on Military Activities

Stoyan Stoyanov

Abstract This article provides a baseline understanding of security environment and military approach to environmental security concepts with a particular focus on understanding the security threats posed by climate change. In this regard, climate change, changing security environment and NATO Strategic Foresight Analysis (SFA) are described briefly. The NATO SFA as a baseline to NATO Long-Term Military Transformation (LTMT), Framework for Future Alliance Operations (FFAO) and NATO Defence Planning Process (NDPP) along with political, human, science and technology and economics/resources poses environment theme in regard to climate change and the Alliance findings related to all of those fields not separately but in their complex interaction.

Keywords NATO • Climate change • Strategic analysis • Future operations

Introduction

In this article, three different topics will be addressed and an overview of the complex interaction between contemporary and future security environmental elements and subsequent NATO military activities to be made. These three topics are as follows: *climate change*, its connection to *security environment* and how it will affect the *Alliance operations* as a part of overall NATO crisis management.

Recently, the world is becoming increasingly more complex, more challenging and less secure, even though globalization and developments in technology are expected to provide ample opportunities for positive developments in health, welfare and security. Increasing interdependency amongst countries has the potential to create stability in the long term.[1] This transition will test the human beings' ability to *adapt* to the challenges of a rapidly changing global security environment.

[1] NATO Strategic Foresight Analysis (Update Report), 2015.

S. Stoyanov (✉)
CMDR COE, Sofia, Bulgaria
e-mail: stoyan.stoyanov@cmdrcoe.org; stoyanov.web@gmail.com

O. Nikolov, S. Veeravalli (eds.), *Implications of Climate Change and Disasters on Military Activities*, NATO Science for Peace and Security Series C: Environmental Security, DOI 10.1007/978-94-024-1071-6_16

Climate Change/Global Warming

In July 2016, CO_2 level was at 404 ppm (it exceeded 400 ppm in March 2015), and, respectively, this particular month was *the warmest in 136 years* of modern record-keeping, according to a monthly analysis of global temperatures by scientists at NASA's Goddard Institute for Space Studies (GISS) in New York. Because the seasonal temperature cycle peaks in July, it means July 2016 also was warmer than any other month on record (Fig. 1). July 2016s temperature was a statistically small 0.1 °C warmer than previous warm Julys in 2015, 2011 and 2009.

The record warm July continued a line of ten consecutive months dating back to October 2015 that have set new monthly high-temperature records. Compared to previous years, the warmer global temperatures last month were most pronounced in the Northern Hemisphere, particularly near the Arctic region. The monthly analysis is assembled from by about 6300 meteorological stations around the world, and the modern temperature record begins around 1880.[2]

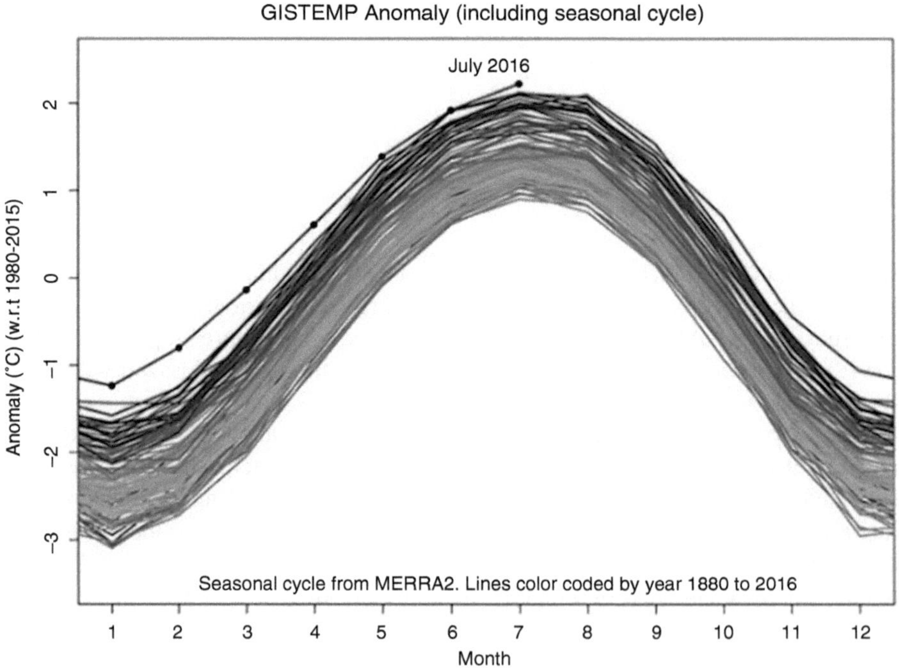

Fig. 1 Recorded monthly temperatures to 2016

[2] http://climate.nasa.gov/news/2479/nasa-analysis-finds-july-2016-is-warmest-on-record/

Ever-Changing Security Environment

Contemporary security environment is a complex alloy of factors, and it is highly influenced by *political governance/regimes, climate change, population growth, globalization and industrialization*. The predicted effects of climate change over the coming decades include *extreme weather events, drought, flooding, sea level rise, retreating glaciers, Arctic sea ice declining, habitat shifts, and the increased spread of life-threatening diseases.* These conditions have the potential to disrupt our way of life and to force changes in the way we keep ourselves safe and secure. The consequences will likely foster political instability where societal demands exceed the capacity of governments to cope. It is worth to be mentioned that if climate change is not the main reason for security disturbances it acts as a *threat multiplier* for instability in some regions of the world. Climate change already amplifies marginal living standards in many countries in Asia, Africa and Middle East. Furthermore, climate change has the potential to result in multiple chronic conditions – complex crises, occurring globally within the same time frame. Economic and environmental conditions in already fragile areas will further erode as food production declines, diseases increase, clean water becomes increasingly scarce and large populations move in search of *resources – food, water, raw materials and energy*. Weakened and failing governments, with an already thin margin for survival, foster the conditions for internal conflicts, extremism and movement towards increased authoritarianism and radical ideologies. Moreover, climate change will add to tensions even in stable regions of the world. In this regard, the Euro-Atlantic Area will continue to experience pressure to accept large numbers of immigrant and refugee populations as drought increases and food production declines in Middle East, Latin America and Africa. When climate change significantly or environmental conditions deteriorate to the point that necessary resources are not available, societies can become stressed, sometimes to the point of collapse.[3]

---

[3] National Security and the Threat of Climate Change, 2007.

NATO Strategic Foresight Analysis

NATO SFA 2015 (SFA 2013 Updated Report) is a product of the Alliance efforts to establish institutional foresight within HQ SACT to provide NATO, national leaders and defence planners with a common perspective of the challenges facing the Alliance in decades to come. The requirement for institutional foresight is affirmed by recent events, including unexpected crises in NATO's immediate vicinity – in the east, Russia's illegal and illegitimate annexation of Crimea and continuing support to separatists in Eastern Ukraine and, in the south, failed or failing states, deepening civil war in Syria and the emergence of Daesh.[4] These emerging issues and the convergence of trends reinforce the need for continuous future horizon scanning in order to support improved decision making. Furthermore, the ability to handle all spectrum of security challenges is vital for the Alliance and requires individual measures at organizational level as well as vision, long-term strategies, program and plans in order to assist the Alliance to perform its core tasks – collective defence, crisis management and cooperate security in a rapidly changing, complex and multipolar future security environment. In this regard, SFA 2015 is an essential component of ACT's Long-Term Military Transformation (LTMT) efforts as it provides input to the Framework for Future Alliance Operations (FFAO) and the NATO Defence Planning Process (NDPP). The SFA 2015 is intended to aid understanding of how current trends could affect the world. These trends could interact or even counteract each other to produce unanticipated consequences. Trends usually converge to create a compound complexity or an instability situation that produces a different or an unanticipated trend. Additionally, the accelerated rate of change within the complex and uncertain future security environment makes reliable anticipation of the future even more difficult. Therefore, the SFA Report(s) provides a shared vision of relevant trend patterns that inform FFAO and support the development of Military Implications (MI) from which defence planners may then derive the *capability requirements* to cope with the complex future security environment and grouping trends into five broad themes: *political*, *human*, *science and technology*, *economics/resources*, *and environment*. In addition, the capability requirements should be taken into account as a part of capability development process which consists of six phases: strategic environment assessment, identifying capability needed, deriving capability requirements, conducting a gap analysis, finding possible solutions (DOTMLPF-I Approach) and applying of the new capability.

Environment Theme

In accordance with SFA 2015, global environmental change and its impacts are becoming readily deceptive and are projected to increase in the future. The main accelerator for that is climate change/global warming. All indicators (IPCC Reports)

[4] Daesh – al-Dawla al-Islamiya fi al-Iraq wa al-Sham.

suggest that the trend is still valid and increasing in regard to severity of extreme weather events and other impacts such as melting polar and glacial ice. However, it is still uncertain what the environmental effects will be by the end of the twenty-first century. This uncertainty is complicated further by the fact that climate change-related environmental effects may have second- or third-order effects on other domains (e.g. political, economic, resources, urbanization and demographics) and may also be affected by future trends in these domains. The severity of this development will potentially increase the number of conflicts based on a mix of different trends and drivers in combination with environmental and climate change. These conflicts may threaten global stability and security and may therefore impact directly or indirectly the members of the Alliance.

On the other hand, natural disaster impacts (e.g. storms, floods, earthquakes) are becoming more devastating. This trend is still valid and increasing in frequency and intensity. It is primarily driven by the construction of infrastructure in disaster-prone areas magnified by the effects of climate and environmental change. Though natural disasters are not of themselves the sole source of conflict or instability, they could deteriorate complex crises. Hereby, the Alliance could be required not only to provide humanitarian assistance and disaster relief support but to plan and to conduct all spectrum of operations described in NATO AJP 3.4. This trend is weaved with the challenges emerging from different phenomena (e.g. industrialization, population growth, urbanization, technological development, climate change, etc.). The underlying drivers will increase and thereby very likely will magnify the destructive effects of natural events. The increasingly interconnected global security system will deteriorate the effects of natural disaster. Although natural disasters can occur anywhere, they will be especially challenging for the political and security system where the social and infrastructural resilience is already weak. This trend might challenge the stability and security in regions within the area of interest of the Alliance. Natural disasters are expected to intensify in frequency and severity as the impacts of climate change increasingly materialize. The impact of natural disasters will become more severe, driven by the growth of megacities in developing countries as rapid urbanization continues. The complex, compound, cascading effects of large-scale natural disasters will be worsened by an increasingly interconnected global economic system, which may amplify the scope of regional disasters towards global impact. Resilience of infrastructure and resources, such as food, water and energy, is increasingly important to mitigate the effects of natural disasters.

Conclusion

According to all of the above-mentioned, security environment will continue to be complex and partially unpredictable. It will be influenced by various factors such as political governance, population growth, globalization, urbanization, technological innovations, industrialization, climate change, etc. All interact in compound manner, and it will be extremely difficult what the security environment will look like in the decades to come. On the other hand, precise foresight is vital for organizations like NATO to develop new or to adjust existing capability in order to be able to cope with those new challenges that will likely arise.

Operation Damayan

Imes Chiu

Abstract Many humanitarian and military leaders noted that civil-military coordination during the Haiyan response in November–December 2013 was some of the best they had seen. Key lessons learned from previous disasters improved the speed and quality of overall coordination. Most notably, personnel with previous disaster response experience who had personal connections with other major players in the relief efforts considerably expedited interagency and transnational relief efforts. The informal professional networks among relief workers built during common training and exercises greatly facilitated the trust needed for effective and efficient cooperation, particularly in the early response phase. The Haiyan relief effort could be equally characterized as an impressive unity of effort despite the lack of standard operating procedures and shared communication platforms. The commitment of foreign assisting actors who came to the aid of the Philippines in Haiyan clearly demonstrated the increasingly globalized nature of disaster response. The ideas, practices, approaches, force capabilities, organizational redesign, and other lessons that came out of the Haiyan experience could potentially be adapted in other disaster scenarios. As global response to large-scale disasters demands a multilayered and multifaceted approach, the need for connective mechanisms becomes critical. In the coming years, the challenge remains to find more innovative ways to increase investment in disaster preparedness and to better integrate and leverage local capabilities and capacities with international response.

Keywords MEB Third Marine Expeditionary Brigade • Armed Forces of the Philippines • Aerial Port of Debarkation • Civil-Military Coordination • Command Operations Center • Civil-Military Operations Center • Disaster Assessment Response Team • Joint Task Force Augmentation Cell • Foreign Disaster Relief • US Marine Forces Pacific • Mutual Defense Treaty • Mission Tasking Matrix • Multinational Coordination Center • Philippines National Disaster Risk Reduction and Management Council • US National Oceanic and Atmospheric Administration • Office of Civil Defense • Office of Foreign Disaster Assistance • Operational Planning Team • Philippine Air Force Tactical Operations Groups • Philippines

I. Chiu (✉)
Daniel K. Inouye Asia-Pacific Center for Security Studies, Honolulu, HI, USA
e-mail: chiui@apcss.org; imes.chiu@gmail.com

O. Nikolov, S. Veeravalli (eds.), *Implications of Climate Change and Disasters on Military Activities*, NATO Science for Peace and Security Series C: Environmental Security, DOI 10.1007/978-94-024-1071-6_17

Disaster Risk Reduction and Management System • Office for the Coordination of Humanitarian Affairs • Operation Damayan • United Nations Disaster Assessment Coordination • United States Agency for International Development • US Government USPACOM United States Pacific Command • World Food Program

Operation Damayan has been widely considered an operational success in terms of efficiency and intra-military communication. The ideas, practices, approaches, force capabilities, organizational redesign, and other lessons that came out of the Haiyan experience could potentially be adapted in other operational environments and disaster scenarios. In highlighting these lessons learned and "best practices" in a narrative context, suggestions introduced focus on what to do rather than what not to do in future relief operations. Many of these "lessons" have been noted before in one form or another, but what is different in this presentation is its application in a catastrophic scenario widely acclaimed by the humanitarian community as one of the best civil-military coordination they have witnessed with 57 other assisting states and 29 foreign militaries.

The best practices in civil-military operational coordination in the Haiyan rapid relief phase include:

Lesson 1: Immediate Request for Assistance and Forward Deployed Assets Saved Lives
Lesson 2: Centralized Planning and Decentralized Execution Facilitated Coordination
Lesson 3: Direct Planning to Ensure Command and Control Is Part of COA Analysis
Lesson 4: Establishment of the International Coordination Team (ICT) Synchronized Effective International Support Through All Phases of USPACOM Operations
Lesson 5: Preplanned Scalable Force Packages Optimize HADR Support

Conclusion

Within 2 weeks, the emergency response phase of the humanitarian crisis in the Philippines in November 2013 was essentially over, and the international community and government agencies were coordinating around shelter and livelihoods. Despite the magnitude of the damage and its wide reach across multiple islands, the commitment of assisting actors who came to the aid of the Philippines clearly demonstrated the increasingly globalized nature of disaster response. This study aims to provide insights into the effectiveness of multinational response to a mega disaster such as Haiyan and hopes to inspire other actors to adapt some of the organizational models and tactical approaches suggested in this study for their particular environments.

Approaches on Building of Resiliency, Mitigation Vulnerability by Radioactive Sources and Management of Radiological Situation in Albania

Luan Qafmolla

Abstract Radiation Protection Commission, Radiation Protection Office, Institute of Applied Nuclear Physics, etc., as the Albanian responsible institutions, have established the regulations system, arranging the preparedness/response for an emergency situation by practices with *s*pent *h*igh *a*ctivity *r*adioactive *s*ource (SHARS). The safety and security assessment at the centralised facility for radioactive waste is performed considering its impact to the workers, public and environment, providing among others the S/S objectives. For a consistent level of institutional control of radioactive wastes discharged or management of SHARS, a policy and strategy for measurement practices and methods to determine radionuclides is implemented already. Safety and security procedures are in operation in IANP facility as a responsible institution for measurement techniques, samples collection and treatment in case of radiological incident/accident happens. To carry out tasks, specialised, qualified emergency working groups-teams operate in emergency location or in laboratory conditions. Teams are established and completed with specialists of IANP, RPO, Geophysical Center and Military Chemistry Division. Those have organised some meetings and exercised/tested emergency plan managed by the American experts. Also, the groups have examined carefully the most important documents for import-export, inventory of sources in use and stored, conditioned and orphan/suspected radioactive sources for 1960–2015.

Keywords Spent radioactive source • Emergency preparedness • Solid waste • Emergency teams

L. Qafmolla (✉)
Institute of Applied Nuclear Physics (IANP), Tirana, Albania
e-mail: l_qafmolla@hotmail.com

O. Nikolov, S. Veeravalli (eds.), *Implications of Climate Change and Disasters on Military Activities*, NATO Science for Peace and Security Series C: Environmental Security, DOI 10.1007/978-94-024-1071-6_18

Introduction

The experience with ionising radiation sources in Albania has shown the necessity of strict actions plan of working rules with *SHARS* sources, in order to avoid any probable radiological incident/accident. Albania has not in operation research reactor (RR) or nuclear power plant (NPP), but in the other side, the country is not far away from other countries that have nuclear power plant, creating probabilities to be under the pressure of a nuclear emergency. The life cycle of radionuclides begins with radioisotopes production at the research reactors, cyclotrons, radiopharmaceutical labs, etc., as the first stage, continuing with their transport at the end users in an application of food irradiation, medical instruments sterilisation, cancer treatment, industrial radiography, scientific research institutions/university and calibration of military devices. The IAEA has categorised radiological/nuclear accidents in five categories, and seeing the ionising radiation sources in use in our country, as well as in geographical position, Albania is included in the third, fourth and fifth categories. Also, the size of the affected/contaminated area is needed to be recognised by the neighbour NPP countries like *Kozloduy in Bulgaria*, *Cernavoda in Romania*, *Paks in Hungary*, *Bohunice in Slovakia*, *Krsko in Croatia–Slovenia*, etc., during nuclear accident (Fig. 1).

In Albania, the General Directorate of Civil Emergency (GDCE) at the Ministry of Interior is responsible for the management of emergency situation including radiological emergency. Since 2001, based at the law No. 8756, the National Operational Civil Emergency Center (NOCEC) was established near GDCE and coordinates/manages the situation of preparedness and response to a civil emergency. The Radiation Protection Commission (RPC) as the national legislative authority and Radiation Protection Office (RPO) as the executive body in cooperation with the Institute of Applied Nuclear Physics (IANP) as the specialised centre coordinate and assist all activities helping GDCE for the managing of radiological emergency situation in Albania. Emergency notification and assistance technical operations manual (ENATOM 2000) are the main conventions and prime legal instrument on early notification of a nuclear accident or radiological emergency. Since 23 August 1957, the Republic of Albania is a member of the IAEA and, since 30 October 2003, has signed the convention on *Early Notification of a Radiological/Nuclear Accident.*

Infrastructure and Status of National Radiation Emergency Plan

GDCE together with RPC operates as a national coordinating authority for preparedness and response to a radiological emergency. RPC prepared the first version of the National Plan for Radiological Emergencies in 2000 [2]. Four years later in 2004, the National Civil Emergency Plan (NCEP) was prepared and approved by

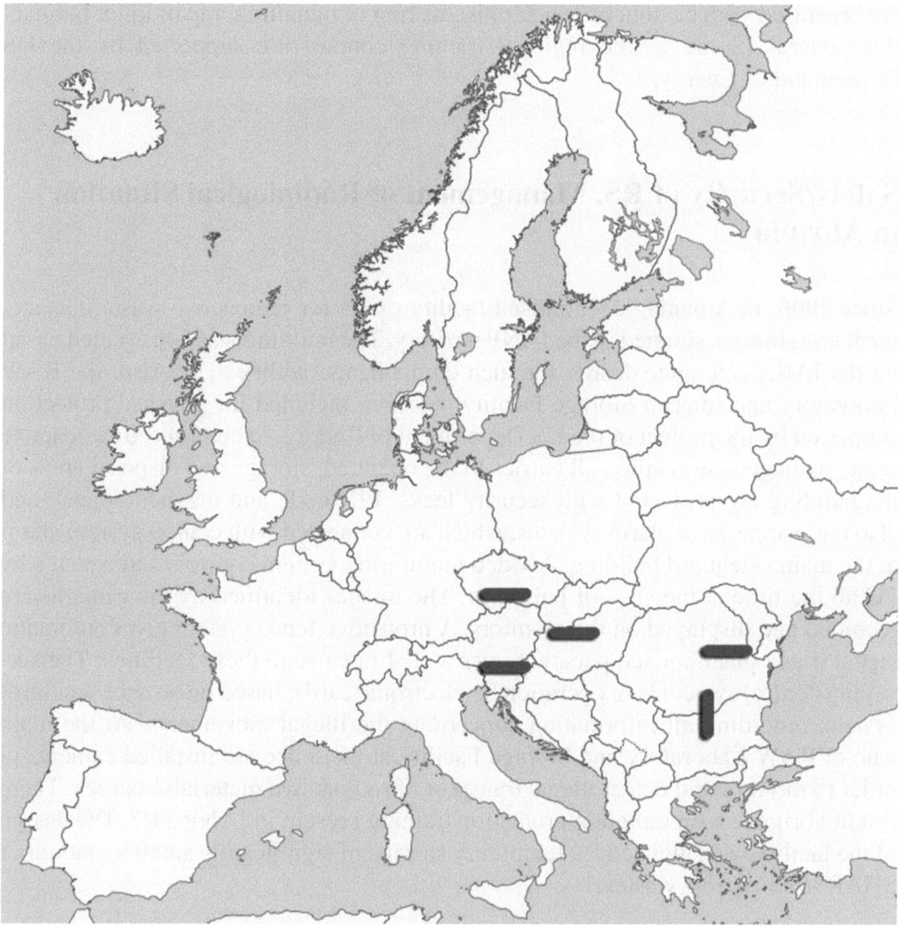

Fig. 1 Neighbour regional nuclear power plants as are categorised by the IAEA [1]

the Council of Ministers. In the same year, RPC issued the revised National Plan for Response to Radiological Emergencies (NPRRE), allocating responsibilities for response to radiological emergency between all parties involved in this plan.

Based in this structure, IANP is a national technical supporting organisation for emergency preparedness and response to radiological emergency. The IANP carries out its jobs of the actual emergency plan by the department of the radiological protection (by both its emergency teams): on-site and laboratory emergency teams. The IANP has hold regularly every 2 years the regional training courses for the staff, which are included in radiological emergency programmes, realising national seminars during 2005–2014 with participations of the personnel from the (**a**) Local and Decentralized Ministry and (**b**) 12 mini-municipalities and General Custom Directory with *main topics*: investigation/detection and found of spent, suspected and orphan radioactive sources used in Albania. Some simulation and table exercises are organised with custom offices for discovering of metallic scrap or illicit radioactive sources. Some of the regional training courses are supported by the US Department of Energy.

Safety/Security of RS, Management of Radiological Situation in Albania

Since 1998, in Albania, a centralised facility exists for radioactive waste management and storage, situated in the IANP territory. The building was constructed based on the IAEA reference design for such components facilities [3]. Also, the RAW Laboratory and Interim Storage Facility has been included for physical protection supported by the project of the US Department of Energy. At centralised radioactive waste management facility, all entries to the operated, storage and disposal areas of the building are protected with security locks, PIN code and magnetic panel, and also these areas have alarm systems, which are connected with central system alarm at the main safeguard building. A video monitoring system controls and watch's by PC all the time at the sites of buildings. The images identified by the cameras are recorded and displayed on the monitors. A protective fence system gives automatic signal if any unauthorised person moves at 3–4 m close to these facilities. The surveying control system is in operation by electronic cards, based on an access control system, recording all information concerning the illegal movements. At the main gate of RAW Laboratory and Storage Facility at the fence are installed cameras in order to monitor and detect illegal transport of radioactive materials/sources. There is a fire brigade with radiation protection training present in IANP 24/7. The design of the facility was intended for temporary storage of significantly smaller amount of SHARS/radioactive wastes.

Evaluation of a Radiological Emergency Situation and Action Plan

The process for establishing and developing of emergency response plan is based on quality assurance programme, which can be verified through the surveying, reviewing, testing and exercising of the plans, procedures and necessary infrastructure. The ability to carry out the required response actions can be evaluated through the above-mentioned actions of the past performance and most commonly through training opportunities, drills and exercises. Training the staff from all relevant emergency response organisations at each level of response, to include drills and exercises, is a key component of a successful emergency preparedness and response programme. They provide an evaluation of continuous improvement programme of the level of compliance with international standards for preparedness in emergency response organisations, tools and procedure [4]s.

Table 1 Study case to discover SRS in Albania for 1960–2015

No	Nuclide	Quantity	Activity mCi	Production year	Users	Imported	Licenced
1	^{226}Ra	3	0,5	1960	IANP/military	Yes	No
2	^{60}Co	1	500	1961	Military/ind.	Yes	No
3	^{226}Ra	2	0,8	1961	Military/hospital	Yes	No
4	^{137}Cs	1	20	2004	Military/IANP	Imported	Yes
5	^{60}Co	1	0,5	2007	Ind./military	Imported	Yes

Study to Discover SRS and Actions/Operations Performed

The IANP and RPO have undertaken a study for spent radiation sources (SRS) situation, to discover abandoned and orphan sources used by the Albanian institutions during 1960–2015 periods. To carry out tasks, a special working group with experts from IANP, RPO, Geophysical Center and Military Divisions was established. An important output of the study was the listing and putting in evidence of orphans/abandoned SRS in country territory studying: invoices, reports, contracts and fundamental register for radioactive sources in use, storage, conditioned and suspected. Since 1971–1995, the database records of radioactive materials and radiation sources are made in IANP by SRS Administrator and Registry programme. Since 1996, with the establishment of RPC, the inventory of SRS and RAW is in duplicate by the RAIS programme at the RPO and IANP. Already, was finished the study related to orphan sources based at bilateral cooperation with RPO (Table 1).

Functions of the Emergency Teams During Emergency Situation

The emergency preparedness and response network system is in total function covering the whole country territory if any probable radiological/nuclear accidents happen. The IANP carries out its jobs during emergency situation by two emergency teams involved at the department of the radiological protection, performing the following tasks: (**a**) to ensure the technical support for the RP staff included in the management of radiological emergency protection to the public and environment by the harmful consequences of emergency situation; (**b**) through the national network established performs emergency monitoring, as well as by both its emergency teams (on-site and laboratory) monitoring, and develop emergency situation, proposing necessary counter-measurements to confront the consequences caused by the accident; (**c**) to realise the samples foodstuff products collection and measurements, as well as measurement of the environmental samples sent in IANP by different organisations, giving the recommendations for the opportunity of their use; (**d**) gives the technical assistance for end users of the SHARS, avoiding consequences for incidents/accident, design their emergency plan ensuring the qualification of the staff for problems related with managing of radiological emergency etc.; (**e**) environmental dose control, air control, contamination control and alpha, gamma spectroscopy and beta total measurements for environmental samples collected, needed to be realised before, during and past a radiological accident.

Conclusions

Harmonisation and combination of actions with local (prefecture level) and national authorities for intervention and protection of the citizens are in accordance with the national/international requirements.

The responsibility for to some past emergency situation coming by stolen and found of SRS shows that arrangements are in place but by the law enforcement and responsible agencies: RPC/RPO and IANP in parallel have indicated their responsibilities solved the emergency situation on the time.

The financial support for intervention or solution of an emergency situation is needed to be covered and ensured by the state budget and other partners' organisation in order for the action plans to have efficacy.

Radiation monitoring level, radionuclide dentification and activity, alpha and gamma spectroscopy and beta total measurements at the site of the centralised facility for samples collected surrounding the territory of the institute are needed to be performed regularly.

References

1. IAEA-TECDOC 718 (1993) A model of national emergency response plan for radiological emergencies
2. IAEA-Safety Series No. 91; Vienna (1989) Emergency planning and preparedness for accident involving radioactive materials used in medicine, industry, research and agriculture
3. Safety Series No. 115, Vienna (1996) International basic safety standards for protections against ionizing radiation and for the safety radiation sources
4. IAEA – TECDOC 953, Vienna (2003) Method for developing arrangements for response to a nuclear or radiological emergency, is updating

Tuning Military Concepts, Doctrine and Force Design to Mitigate Human Security Challenges Arising from Global Environmental Change

Jeffrey A. Andrews and Etienne F. van Blerk

Abstract The South African Department of Defence considers the effects of climate change within a broader context of global environmental change, along with such drivers as freshwater decline; land degradation and desertification; stratospheric ozone depletion; and loss of biodiversity and ecosystem functioning. All of these elements converge to contribute significantly to contemporary environment-related threats to human security, more so in the developing world. Critics from both civil society and the military question the legitimacy of military involvement in government's response to these, mostly non-military threats. These questions are placated in this rationale, which examines the nature of human security as it interfaces with the environmental security-related system, the corresponding institutionalised response on the part of the international community and the manner in which these initiatives ultimately involve conventional military capabilities. Such engagement demands specific mission capacity in addition to mission sustainability in the face of global environmental change.

Keywords Human security • Environmental security-related system • Global environmental change • Sustainable development • Military-integrated environmental management • Environmental legislation • Encroachment • Environmental hazards • Environmental conflict • Resilience

Introduction

The confluence of various developments in international relations in recent times has brought on a dramatically altered security agenda all but devoid of the need for a conventional war force. In this time, the emergence of various global

J.A. Andrews (✉)
United States Africa Command, Stuttgart, Germany
e-mail: jeffrey.a.andrews.civ@mail.mil; four.andrews@gmail.com

E.F. van Blerk
Lieutenant Colonel at South African Air Force, Pretoria, South Africa
e-mail: etiennef.vanblerk@gmail.com

O. Nikolov, S. Veeravalli (eds.), *Implications of Climate Change and Disasters on Military Activities*, NATO Science for Peace and Security Series C: Environmental Security, DOI 10.1007/978-94-024-1071-6_19

environment-related threats arising from such phenomena as climate change among others, mostly associated with intensified incidence of natural disaster and human distress, has added a new dimension of non-military threats to the security agenda of the twenty-first century.

Other than deployment in wars, military forces are unique in their capacity to muster the command and control as well as logistical capabilities, with a specially trained and disciplined workforce, needed to provide relief in the chaotic conditions resulting from man-made or natural disasters. The now-altered security agenda where military forces are more likely to be deployed in operations other than war, such as disaster relief, has occasioned widespread debate, also in South Africa, into the intricacies of military assistance to civil authorities [11]. Operations other than war in this domain, furthermore, have drawn criticism from civil society for so-called militarisation of environment-related threats such as climate change, supposedly both to legitimise the issue and to "moralise the military" in a ploy to leverage resources [8]. Staunch military proponents, on the other hand, may view any substantial shift to this kind of force employment as "mission creep", which is deemed to detract from primacy of a constitutional military mandate to defend and protect sovereign territory [11].

The recent developments in international relations and emergence of environment-related threats with global reach are however likely to overshadow the raging debate with an unavoidable rationale for a renewed perspective on military concepts, doctrine and force design tuned to human security challenges arising from global environmental change.

Human Security

Until the end of the Cold War, signified by the fall of the Berlin Wall in 1989, a prevailing concept of state security had been determined largely by military threats. A debate about the true meaning and re-conceptualisation of security that had first opened in the 1970s gained thrust during the 1980s to give shape to the concept of human security [12]. To this day, the concept of human security remains in force as a more comprehensive substitute in the post-Cold War era to the once narrow and single-minded concern with security of the state.

Human security, in this sense, considers not only military determination of threats but also politically, economically, socially and environmentally motivated threats. These five sources of threats or security-related systems, furthermore, are encountered at the level of an individual, that of the state and at an international level. The sovereign state according to Buzan [2], however, remains at the centre as an important and effective provider of security. Strong states would be better equipped accordingly to deliver human security than would weak states. Booth [1] argues that apart from the state, non-state actors such as individuals, ethnic or cultural groups, regional economic blocs, multinational corporations, non-governmental organisations and the like would strongly influence efforts on the part of any sovereign state to render human security.

The United Nations [21] ultimately defines human security as "...protecting fundamental freedoms – freedoms that are the essence of life; protecting people from severe and widespread threats and situations; using processes that build on people's strengths and aspirations; creating political, social, environmental, economic, military and cultural systems that together give people building blocks for survival, livelihood and dignity...".

This contemporary premise for human security was already embedded in the South African government's ambition for its armed forces in the first edition of Defence Review [3] and is reaffirmed in a more recent revision [5] as a guide to developing core functions and force structure of the South African National Defence Force (SANDF).

Environmental Security-Related System

Local environmental conditions such as droughts, floods, storms or epidemic disease and other environmental constraints and their natural variability have always impacted the security of individuals or societies as causes of mortality, morbidity and social disruption [13]. Modern societies with their technologies, trade, industrialisation, use of fossil fuels and higher levels of social organisation have weakened the effects of local environmental constraints on human security.

Global economic activity of production, distribution, trade and consumption has burgeoned since the Industrial Revolution and rise of today's trading nation-state. In this time moreover, the global population has grown to more than seven billion, while most people now live longer, consume more and are better educated than in previous generations. The once-localised environmental risks to human security are however all but gone. Instead, the staggering scale of global economic activity is saturating the atmosphere with greenhouse gases to accelerate global warming, sea level rise and climate change, polluting available freshwater in the face of increased demand for this resource, stripping land cover for extensive territories to degrade to irreparable wasteland, emitting industrial chemicals that erode the stratospheric ozone shielding the planet from harmful solar radiation, exploiting biodiversity to the point of collapse of staple (e.g. fisheries) and high-net-value (e.g. rhino) biological resources and destroying ecosystems needed to deliver life-supporting natural utilities and recovery systems. This regression has assumed global proportions insofar as there are linkages between environmental changes and social consequences across distant settings, groups and time horizons in the world today.

Proctor [15] therefore views global environmental change as the set of biophysical transformations of land, oceans and the atmosphere, driven by an interwoven system of human and natural processes that are intimately connected with processes of socio-economic and cultural globalisation. Matthew and Fraser [14] bring global environmental change into context with human security arguing that global environmental change can be a direct source of insecurity.

The prevailing forms of global environmental change cited by various authors can be grouped into five main drivers [16]: (1) climate change, climate variability or climate disruption, (2) freshwater decline, (3) desertification and land degradation, (4) stratospheric ozone depletion and (5) loss of biodiversity and ecosystem functioning.

Each of these drivers of global environmental change deserves treatment in their own right. All of these drivers, however, share interrelationships among each other; the adverse trans-boundary effects of any single one of these groupings, being either intensified by another; or serving to worsen the effects associated with any of the other groupings. The South African Department of Defence therefore assumes this cumulative perspective of global environmental change in its appraisal of the environmental security-related system that is incorporated into the most recent revision of the South African Defence Review [5].

Worldwide Institutional Response

Global environmental change, as the term would suggest, demands international collaboration in efforts to stem its causes and to mitigate its effects. The first most significant international response in this respect came about in 1972 when the United Nations (UN) General Assembly convened its Conference on the Human Environment in Stockholm, Sweden, to deliver a declaration with 26 principles concerning the environment and development, an action plan and a resolution [9]. Following this continuum by 1983, the UN General Assembly established the Brundtland Commission. The commission, formally known as the World Commission on Environment and Development, produced "Our Common Future" [18], a report in which countries are rallied to work together in pursuit of sustainable development. The commission further called for a meeting of world leadership where international initiatives and goals could be mapped out. The result was the Earth Summit, convened in Rio de Janeiro, Brazil, in 1992 to give shape to a comprehensive plan of action, Agenda 21 [19], which was to guide global action nationally and locally in order to render life on Earth more sustainable into the future.

The Johannesburg Declaration on Sustainable Development [20] was adopted at the World Summit on Sustainable Development held in Johannesburg, South Africa, in 2002. This declaration builds on the earlier international commitments to sustainable development, but adds to these in sponsoring multilateralism in mapping the way ahead. As an international agreement, it focuses particularly on the worldwide conditions that pose severe threats to the sustainable development of people in the face of global environmental change.

Cabinet approved South Africa's National Strategy for Sustainable Development (NSSD) in 2011 [6] as a commitment to furthering the effective stewardship of the country's natural, social and economic capital. Contemporary developments, such as the lingering global financial crisis and the international focus on climate change and emphasis on "green economy" (UNEP's "Global Green New Deal"), were all factored into formulating the NSSD Action Plan. The NSSD Action Plan draws upon the five strategic priorities of South Africa's national sustainable development framework to include (1) climate change response, (2) greening the economy, (3) building sustainable communities, (4) sustaining ecosystems and efficient resource use and (5) integrated planning and implementation systems.

Numerous international sustainable development-related institutions, agreements, treaties, customs and general principles have indeed emerged since 1972 aimed at either stemming the causes of global environmental change or mitigating its effects on the economic growth, social equity and environmental integrity on which nations depend. Member states that are party to sustainable development-related international conventions ratify these worldwide initiatives into domestic law. This represents a significant step on the part of governments towards mainstreaming of sustainable development as a countermeasure to the ravages of global environmental change.

Implications for the Armed Forces in South Africa

Diverse ministries and lead agencies in national, provincial and local government in South Africa are mandated and resourced to generate domestic sustainable development-related policies, to deliver associated services and to enforce policies. The Department of Defence or its constituents do not resort among the more apparent lead agents in the government for the implementation of sustainable development in South Africa. What cannot be denied though is the role of this department both as a stakeholder in co-operative governance for sustainable development [4] and the potential vulnerability of military mission sustainability to the effects of global environmental change. As such, the Department of Defence [5] aspires, first and foremost, to build resilience to withstand the effects of global environmental change on the military footprint and, secondly, to support the collective efforts of civil authorities in strengthening resilience of the society it serves. Response on the part of the South African armed forces in its approach to the implications of global environmental change is therefore focused in the following two areas:

1. The first focus area of household military-integrated environmental management aims to manage all facets of the military footprint to ensure long-term sustainability of both the military mission requirement *and* the receiving environment. The approach here is based on the following two strategic imperatives:

   - *Legislation.* Emergent sustainable development-related legislation aimed at stemming the causes or mitigating the effects of global environmental change has escalated requirements for legal compliance, also by the armed forces. This has necessitated establishment of a professional and well-constituted, organic body of military environmental management capacity since 1978, thereby enabling the SANDF to exercise mandatory environmental responsibilities in the public interest while maintaining military mission sustainability interests of its own.

- *Encroachment.* Global environmental change has both a direct and indirect impact on sustaining military capabilities and force dispersion. Biophysical effects associated with global environmental change affect military operations, assets, personnel and infrastructure. Rigorous contingency planning is therefore aimed at improving resilience of military facilities in the most probable biophysical effect scenarios (e.g. disruption of bulk water supply). National strategic priorities for sustainable development, on the other hand, have inspired, among others, the expansion of protected areas, proliferation of renewable energy schemes and patterns of human settlement that inadvertently converge on military spatial requirements. Mechanisms for structured, proactive civil-military co-operation are imperative here, to resolve competing civil-military spatial requirements [7].

2. The second focus area of environmental security aims to explore threats that environmental events and trends pose to individuals, communities or nations with a focus on the impact of human conflict and international relations on the environment or on how environmental problems cross state borders. The approach in this instance is based on the following two strategic imperatives:

    - *Environmental hazards.* Civil authorities call for the mobilisation of military assets in dealing with more frequent national or trans-boundary hazards arising from global environmental change. Global environmental change, increased human population, dispersion and vulnerability of communities have given rise to more frequent natural disasters worldwide [22]. Drought is deemed the most significant of natural disasters occurring in South Africa, if not for its infrequent occurrence or slow onset, certainly then for its intensity and devastating long-term effects on entire communities as well as a developing economy. Droughts in the region are accompanied moreover by other natural hazards of flash floods and wildfire. To this effect, the South African Air Force [17] issued a standing directive for the implementation of drought measures to exact preparations for diverse modes of responsiveness as an organisation.
    - *New threat patterns.* The security agenda arising from global environmental change will determine adjustments to existing concepts, doctrine and force design. Homer-Dixon [10] describes threat development towards various forms of environmental conflict that can be useful in identifying indicators for early warning and determining the mode and intensity of military intervention for the purposes of higher-order joint force employment planning.

Conclusion

Rapid maturation of globalised human economic activity mostly over the past 150 years has accelerated development and improved the quality of life for most people in the world. The rampant and insatiable rate of resource extraction, production, distribution, trade and consumption demanded by such activity has however come at a cost. Widespread and diverse environmental changes are occurring across the planet to the extent that these are threatening the trends of economic activity that contribute to these changes. As an added result of these changes, the livelihoods and quality of life of people across the world is now becoming more insecure in many ways.

Two significant developments, the evolution of an all-encompassing concept of human security and that of sustainable development, are being harnessed by the international community as an institutionalised resolve to a compounded problem, which forms a part of an intricate global security agenda. A favourable prognosis, however, demands cross-sectoral collaboration both among nations and within. As an organ of state, therefore, the involvement of military forces in the governance of sustainable development for human security is inevitable.

Although global environmental change is characterised by unconventional and non-military threats to human security, standing conventional military forces still possess unique capabilities and qualities that are ideal for lending support to civil authorities in their efforts to stem the causes and mitigate the effects of global environmental change. Co-operation of this kind may therefore negate any apparent necessity for adjustments to the force design of conventional military forces. Instead, it would imply a shift in the traditional emphasis on defending and protecting sovereign territory against foreign military incursion and more towards employing standing capabilities to augment the capacity of mandated civil authorities in a collaborative effort of building robust institutions, equitable societies and resilient communities, in a word, a strong state. Doctrinal development would further forge and hone the ability of military forces and the practiced confidence of command to render such support constructively within the bounds of sound civil-military relations and co-operative governance.

Finally, global environmental change brings with it certain threats to military mission sustainability. Military forces are therefore compelled to appraise these threats and to ensure resilience of military facilities and capabilities in order to maintain the levels of readiness and capacity to act in times of crises.

References

1. Booth K (1994) A security regime in Southern Africa: theoretical considerations, Southern African perspectives No. 30, CSAS
2. Buzan B (1991) People, states and fear: an agenda for International Security Studies in the post-cold war era. ECPR Press, Colchester
3. Department of Defence (1998) South African defence review. 1 Military Printing Regiment, Pretoria, Republic of South Africa
4. Department of Defence (2001) First edition environmental implementation plan for defence. General Notice No. 249 under the National Environmental Management Act (Act No. 107 of 1998), Government Gazette No. 22022 of 16 February, Pretoria, Republic of South Africa

5. Department of Defence (2013) South African defence review. 1 Military printing regiment, Pretoria, Republic of South Africa
6. Department of Environmental Affairs (2010) Draft National strategy for sustainable development and action plan 2010–2014. General Notice No. 393, Government Gazette No. 33184 of 14 May, Pretoria, Republic of South Africa
7. Environmental Security Working Group (2010) Guidebook on outreach for mission sustainability. A joint United States – Republic of South Africa Environmental Security Working Group Project, Publication ESWG/009, Washington DC, United States of America
8. Gilbert E (2012) The militarization of climate change. ACME: an International E-Journal for geographies, vol 11, No. 1, Canadian Studies & Geography, University of Toronto, Toronto, pp 1–14
9. Handl G (2012) Declaration of the United Nations Conference on the Human Environment (Stockholm Declaration), 1972 and the Rio Declaration on Environment and Development, 1992, United Nations Audiovisual Library of International Law
10. Homer-Dixon K (1991) On the threshold: environmental changes as causes of acute conflict. Int Secur 16:76–116
11. Kruys GPH (2009) Military assistance to civil authorities: the case of South Africa. ISSUP Bulletin No. 5, 2009, Institute for Strategic Studies, University of Pretoria, Republic of South Africa, pp. 1–16
12. Naidoo S (2001) A theoretical conceptualisation of human security. Peace, human security and conflict prevention in Africa: Proceedings of the UNESCO-ISS Expert Meeting, Pretoria, Republic of South Africa
13. Matthew RA, Bartnett J, McDonald B, O'Brien KL (2010) Global environmental change and human security. Massachusetts Institute of Technology, The MIT Press, Cambridge, MA, p 3
14. Matthew RA, Fraser L (2002) Global environmental change and human security: conceptual and theoretical issues. Global Environmental Change and Human Security Program, University of California, Irvine, p 13
15. Proctor JD (1998) The meaning of global environmental change: re-theorizing culture in human dimensions research. Global Environ Change, Elsevier Science Ltd, Great Britain 8(3):238
16. South African Air Force (2009) Department of Defence Environmental Analysis: South African Air Force Contribution. Report CAF/DLOG/R/401/1/3/9 dated 23 October 2009, Directorate Logistics, Air Force Office, Pretoria, Republic of South Africa
17. South African Air Force (2016). Drought measures for implementation throughout the South African Air Force. Chief of Air Staff Operations Instruction 02/2016 CAS OPS/R/304/1 dated 23 February, Chief of Air Staff Operations, Air Force Headquarters, Pretoria, Republic of South Africa
18. United Nations (1987) Report of the world commission on environment and development: our common future
19. United Nations (1992) United Nations Conference on Environment & Development: Rio de Janeiro, Brazil, 3 to 14 June 1992: Agenda 21, United Nations Division for Sustainable Development
20. United Nations (2002) Johannesburg declaration on sustainable development. A/CONF. 199/20, Chapter 1, Resolution 1, Johannesburg, Republic of South Africa
21. United Nations (2009) Human security in theory and practice: an overview of the United Nations trust fund for human security. United Nations Human Security Unit, New York, p 5
22. Van Blerk EF (2007) Natural disasters. Enviropaedia. Be the change Edition, Eco-Logic, Simonstown, p 198

Environmental Security and Climate Change: "A Security Assessment of the Balkan Region"

Wendell C. King, Kosta Delev, and Kuman Gerovski

Abstract This presentation provided a baseline understanding of the concepts of environmental security with a particular focus on understanding the security threats posed by the key environmental security drivers of climate change and water as a scarce resource. The presentation defined environmental security as a process where human-induced environmental changes have the potential to deny people access to their basic human needs and thereby create security threats. Case studies of Afghanistan, the Nile River watershed, and the Tibetan Plateau were used to illustrate how environmental change with population growth creates security threats and can lead to failed states or even open conflict. A special emphasis was given to examining the security threats resulting from the key climate change drivers. The presentation closes with presenting key findings from master's theses from two Bulgarian officers who conducted environmental security and climate change studies for the Balkan Region.

Keywords Environmental security • Climate change • Water scarcity • Failed states

Introduction

This presentation serves as an overview of the basic principles of environmental security analysis in order to establish a common understanding across the broad range of backgrounds of the conference attendees. It started by defining environmental security as an academic discipline. It also reviews the scientific work of the UN Intergovernmental Panel on Climate Change (IPCC). This work then allowed a detailed examination of the climate change impacts that have security and defense

W.C. King (✉)
U.S. Army Command and General Staff College, Fort Leavenworth, KS, USA
e-mail: wendell.c.king.civ@mail.mil; wendellcking@yahoo.com

K. Delev
NATO Force Integration Unit-Bulgaria, Sofia, Bulgaria
e-mail: Kosta.DELEV@bgr.nfiu.nato.int; kostadinkov@gmail.com

K. Gerovski
Bulgarian Army, Sofia, Bulgaria

O. Nikolov, S. Veeravalli (eds.), *Implications of Climate Change and Disasters on Military Activities*, NATO Science for Peace and Security Series C: Environmental Security, DOI 10.1007/978-94-024-1071-6_20

implications. The final element of the presentation was a presentation of an environmental security analysis for the Balkan Region and a threat assessment of the security implications of climate change in the Balkans.

Key takeaways – Environmental security is a process where human-driven environmental change threatens the security and stability of a region by denying people access to their basic human needs. Environmental security threats include loss of arable lands, water scarcity, deforestation, desertification, and climate change. Population growth is an independent variable that has a major impact on environmental security.

Climate change is a world-scale issue as documented in the IPCC 5th Assessment Report [1]. The adverse impacts of climate change are already occurring and will grow without major efforts to mitigate the rate of greenhouse gas emissions. People and nations will have to adapt to already existing impacts of climate change. The climate impacts and their associated defense and security impacts are shown in the table [2]. Each has impacts on military operations and defense planning. NATO and the countries in the Balkan Region must act now to address these issues.

The largest environmental security issue for the Balkan Region will be water resource management [3]. This will result in both periods of heavy flooding and also drought and reduced water for agricultural uses. In addition to the water resources issues, climate change will present a variety of other security threats to the region. Refugee populations from other regions impacted by climate change will be a major problem for the Balkan Region [4]. Internal to the region, human security issues from disease and extreme heat and weather will stress the population and damage the countries economically. In response, the military forces of the region will need to prepare for these threats. This may include changes to force structure and equipment, better contingency planning for disasters, and more coordination with civilian agencies. A key will be building cooperation between the nations of the Balkan Region.

Conclusion

Climate change and other environmental security issues will pose significant challenges to the countries of the Balkans and their military forces. The external refugee migrants displaced by climate change will be a major external threat [4]. Internally, countries will be threatened by extreme weather, loss of arable lands, reduced water supplies, and more. These issues will create economic problems and human health and safety concerns which will at times overwhelm civilian response capacity. Military forces need to prepare for these new missions. Force structure, training, and equipment must be assessed against these new challenges to peace and security in the region. The most important change that must occur is that the countries of the region must develop plans and coordinate responses to the disasters that are to come in order to limit the human impacts of climate change (Table 1).

Table 1 Major impacts of climate change

Table references specific sections in the IPCC Summary for Policy Makers, 2014		
Climate-related driver	Key impacts	Security and defense impact
Temperature warming: 0.85 ° C in 2012, 1.0–3.7 °C by 2100 (WGI, SPM-3)	Increase of disease (vector and water borne), stress on water resources, loss of arable lands, reduced food production, increase in salinity, degrading of coral reefs, loss of fish stock and livelihoods	Increase of humanitarian support missions, refugee support, medical resources to respond to epidemic disease, potential for conflict
Extreme temperature: highest in Asia, Europe, Australia, (WGI, SPM-15, 23)	Increased mortality and health and well-being issues, stress water resources, reduced crop production	Medical logistics support, increase of humanitarian support missions, security operations (ops), and potential for conflict
Drying trend: global, highest in midlatitudes (WGI, SPM-23)	Food security threats, water resource stress	Support migrations, humanitarian ops, potential for conflict
Extreme precipitation: highest in midlatitudes and wet tropics by 2100 (WGI, SPM-16)	Flood damage to infrastructure, loss of life, increased infectious and vector-borne disease	Increase of humanitarian support missions, large-scale logistics support, medical ops in response to epidemic disease, security ops
Precipitation: More in the high latitudes and at the equator. Drier in midlatitudes and subtropics (WGI, SPM-17)	Water resource stress, loss of arable land, public health issues, water quality degradation	Increase of humanitarian support missions, logistics support, medical support to respond to epidemic disease, security ops, potential for conflict, engineering support
Snow and ice cover: Ice – 15–85 % reduction by 2100. Snow – 7–25% loss by 2100 (WGI, SPM-17)	Loss of snow and ice stresses water resources, increased rate of warming, flooding, and droughts	Increase of humanitarian support missions, large-scale logistics support, medical resources to respond to epidemic disease, border security ops
Damaging cyclone: most likely in western North Pacific and North Atlantic (WGI, SPM-23)	Loss of life and property damage, extreme flooding, increased disease following disaster	Increase of humanitarian support missions, security ops, engineering reconstruction support, disaster medical relief, logistics support
Sea level: 0.19 M in 2010, 0.4–63 by 2100 (WGI, SPM-18)	Flooding/property damage, loss of coastal and island settlements, reduced food production, water quality damage	Refugee support, large-scale logistics support, security ops

References

1. IPCC (2013) Summary for policymakers. In: Stocker TF, Qin D, Plattner G-K, Tignor M, Allen SK, Boschung J, Nauels A, Xia Y, Bex V, Midgley PM (eds) Climate change 2013: the physical science basis. Contribution of Working Group I to the Fifth Assessment Report of the Intergovernmental Panel on Climate Change. Cambridge University Press, Cambridge

2. King WC, Morisetti N, Singh J (2014). Strategic defense impacts of climate change. OSCE Security Days Conference, Vienna
3. Delev K (2015) Climate change and poor water resource management will have serious security implications in the Balkan Peninsula, Masters thesis at US Army Command and General Staff College
4. Gerovski K (2016) Climate change impact on the Southeastern Europe Security Environment and the increasing role of the Bulgarian Army as the World Warms. Masters thesis at US Army Command and General Staff College

Climate Change: Fundamentals, Agroclimatic Conditions in Bulgaria, and Resilience Agriculture Through Adaptation

Valentin Kazandjiev

Abstract The important factors for the agrarian output in Bulgaria are only thermal and water probability. From the two factors, the component related to soil moisture is more limited. As well water and temperature probabilities in the agrarian output are estimated through sums of temperatures and rainfalls or by derivative indicators (most frequently named as coefficients or indices).

The heat conditions and the heat resources are specified by the continuousness of the vegetative period. Duration of vegetative season is limited for each type of plant, between the spring and autumn steady pass of air temperature across the biological minimum. For the agricultural crops in Bulgaria, the three biological minimums in 5 °C are taken for wheat and barley, oat, pea, and lentil; in 10 °C for sunflower, corn, haricot, and soybean; and in 15 °C for the cotton, vegetables, and other spring cultures.

The cold and warm period duration are mutually related characteristics. The first period defines the number of days with the snowfall and days with the snow cover that are the basis in the formation of soil moisture reserves after the spring snow melt. Definition of the regions with temperature stress conditions during vegetative season is one of the most important parameters of agroclimatic conditions. The values indicating for the limitations are one or more periods from at least 10 consecutive days with maximal air temperature over 35 °C. More from the agricultures, character for the moderate continental climatic zone are developed normally under temperatures 25–28 °C. Temperatures over 28 °C are ballast slowing the growth and destroying plants due to the heat tension. The component, limiting in greatest degree growth, development and formation of yields from the agricultural crops are the conditions of moisturizing, present trough atmospheric and soil moisture. The most apparent indicator is the year sum of the rains or their sum by the periods with the average daily temperatures of over 5 and 10 °C. Cross correlation matrix between the meteorological elements from which evapotranspiration depends – temperature, relative air humidity, wind speed, and the vapor pressure deficit – is present.

V. Kazandjiev (✉)
National Institute of Meteorology and Hydrology – BAS, Sofia, Bulgaria
e-mail: valentin.kazandjiev@meteo.bg; valentin.kazandjiev@gmail.com

O. Nikolov, S. Veeravalli (eds.), *Implications of Climate Change and Disasters on Military Activities*, NATO Science for Peace and Security Series C: Environmental Security, DOI 10.1007/978-94-024-1071-6_21

The data about the limitations, emergent from the soil moisture lack, to the base of the existing agrometeorological data are present. Values of the relation between real and potential evapotranspiration were calculated for potential vegetative period which is divided up to the two subperiods, March to June, the period of formation outputs from wintering cultures, and July to August, the period of formation outputs from the spring cultures.

In the 1980s and 1990s, science led debates "for" and "against" climate change. During this time they published dozens of monographs and among them are Sir John Houghton's *Global Warming: The Complete Briefing* and John T. Hardy's *Climate Change: Causes, Effects, and Solutions*. The first of them was translated into Bulgarian by the author of this paper and published in 1996 by the academic publishing house of Prof. M. Drinov. Of course, they published numerous other studies and hundreds of articles, reports, and messages (Olmstead, Rhode, Creating abundance: biological innovation and American Agricultural Development. Cambridge University Press, 2008; Croitoru et al, Glob Planet Change 102:10–19, 2013; Rosenzweig, Hillel, Climate change and the global harvest: potential impacts of the greenhouse effect on agriculture. Oxford University Press, 1998; Georgieva, Kazandjiev, Sci Pares Ser A Agron LVI:459–467, 2013; Georgieva et al, Europa XXI 29:43–58, 2015; Kazandjiev, Peev, Prerequisites for disaster by natural weather phenomena and processes, reports first scientific-practical conference on Emergency Management and Civil Protection, Sofia, Bulgarian Academy of Sciences 10.11.2005, pp 186–193 (in Bulgarian), 2005d; Kazandjiev, Agroclimatic resources and definition of less favored areas at the beginning of XXI century in Bulgaria, Conference "Global Environmental Change – Challenges to Science and Society in Southwestern Europe." CD version, 2008a; Rattan et al, Climate change and global food security, CRC, 2005; Roumenina et al, Int J Remote Sens 34(8):2888–2904, 2013; Pritchard, Amthor, Crops and environmental changes. Haworth Press (US), 2005; Simeonov, Georgiev, Atmos Res 57:187–199, 2001; Sivakumar et al, Natural disasters and extreme events in agriculture. Springer, 367 pp, 2005; Slavov, Relationship between climate change and desertification. Problems of land degradation and combating desertification. UN str.42-48 (in Bulgarian), 1998; Slavov, Alexandrov Drought Netw News 5(2):12–15, 1993).

Today science has a lot of evidence in favor of climate change. But now science nationally and globally faces new questions:

- How far will climate change reach?
- How will the various sectors of the economy adapt to change?
- How will agriculture in particular adapt to climate change?
- What must the action plan 2030–2050 contain?

The purpose of this paper is to plot a strategy for the adaptation of agriculture in Bulgaria to climatic change. This will establish the vulnerability of the main types of crops to climate change and will define criteria for extreme meteorological phenomena and processes of agro-meteorological point of view. The team will assess the risk of dangerous agriculture phenomena and combinations thereof, through probabilistic and statistical research. Also we will present indices that can be used as indicators for proof of climate change. As a result, they will identify adaptation measures by regions and types of cultures and develop a strategy for adaptation of Bulgarian agriculture to changing environmental conditions.

Keywords Climate change • Agriculture • Resilience • Adaptation • Security • Disaster response

Introduction

Today there is no doubt in existence of climate change, because only a few years before, the skeptics maintain that these changes are a fiction created from some politics with which climate changes explain some of their failures. Now the already actual questions are:

- What is magnitude of changes?
- How can one characterize, explain, and present changes?
- What is the trend of changes?
- How do the changes impact on agriculture and other sectors?
- Which is the way to mitigate effect of changes and develop sustainable agriculture?
- What is the place of zoning to optimize using agroclimatic resources?

Aims of this investigation are to find the dynamics and trend of temperature time series, spatial distribution of average, and maximal and minimal annual temperatures; to find the dynamics and trend of rainfall time series and their spatial distribution; to investigate stress conditions following from high temperatures and low temperatures or insufficient moisturizing conditions; and finally to investigate necessity for zoning of agroclimatic resources in Bulgaria in dependence of mentioned parameters. All these targets are realized using data from climatic observations in 60 stations part of meteorological network of National Institute of Meteorology and Hydrology of Bulgarian Academy of Sciences (NIMH-BAS). Used experimental data was for last 35 years, i.e., 1971–2005 for next meteorological elements and phenomena minimum and maximum temperatures, rainfalls and their derivatives. relative air humidity, vapor pressure, sunshine duration and number of days with rain.

One of the sectors that is significantly affected by climate change is agriculture, which provides food supplies to the population without food. Sustainable development of this sector in a changing climate is essential to national security. From 1970 we have observed an increasing frequency and intensity of dangerous weather conditions for agriculture such as heavy rains and flooding, hail storms, high winds, drought, heat waves, and frost and change in seasonality of weather, sudden warming in winter, and cooling in summer.

Thermal Conditions

Dynamics of average annual temperatures for the period 1900–2010, their difference in comparison with 30-year period 1961–1990, and their moving average to these differences and trend of change are present on Fig. 1. For investigated period average annual temperatures are increased from 11 to 13 °C or with ~1.5 °C in relation with 1961–1990. In some years differences are with negative sign – 1972, 1976, 1980, 1990, 1992, 1995, 1996, and 1997 [6, 7, 10–17]. During all other years, differences are positive, i.e., warming is observed. How is this warming distributed on the territory of Bulgaria? Data for the last 20-years period was drawn in a map – Fig. 2, with spatial distribution of average annual temperatures over the country. The most affected from warming process are the territories of central and eastern part of Bulgaria but too strong in south Central Bulgaria. **Peculiarities of average temperature spatial distribution are connected with type of climates in these parts of the country – in north, moderate continental and, in south, transitional Mediterranean.**

Absolute maximum of temperature for this period is measured on 05 July 2000 in Yambol, 44.4 °C. Figure 3 presents spatial distribution of annual average maximum temperatures. Relatively lower are the absolute maximums in the northeast and central west parts of Bulgaria.

The coldest point in the country is situated in Kneja, northwest of Bulgaria, where on 13 January 1985, absolute minimal temperature of –29.3 °C is measured. Figure 4 presents spatial distribution of annual average of minimum temperatures. As a result climatic amplitude of temperatures during the last 30

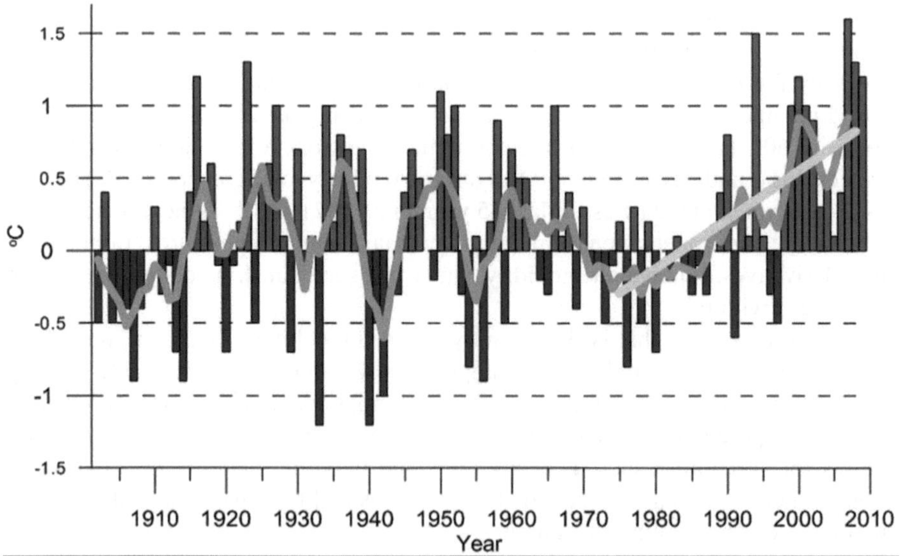

Fig. 1 Anomalies of air temperature in Bulgaria in comparison with the period 1900–2010 [2, 3]

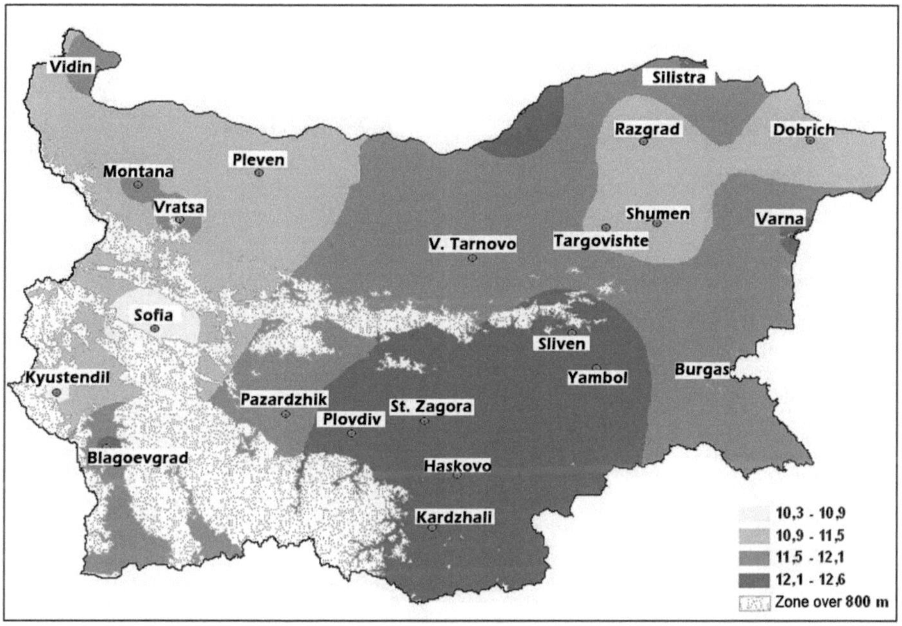

Fig. 2 Spatial presentation of average annual temperatures within the country in (°C) for the period 1996–2015

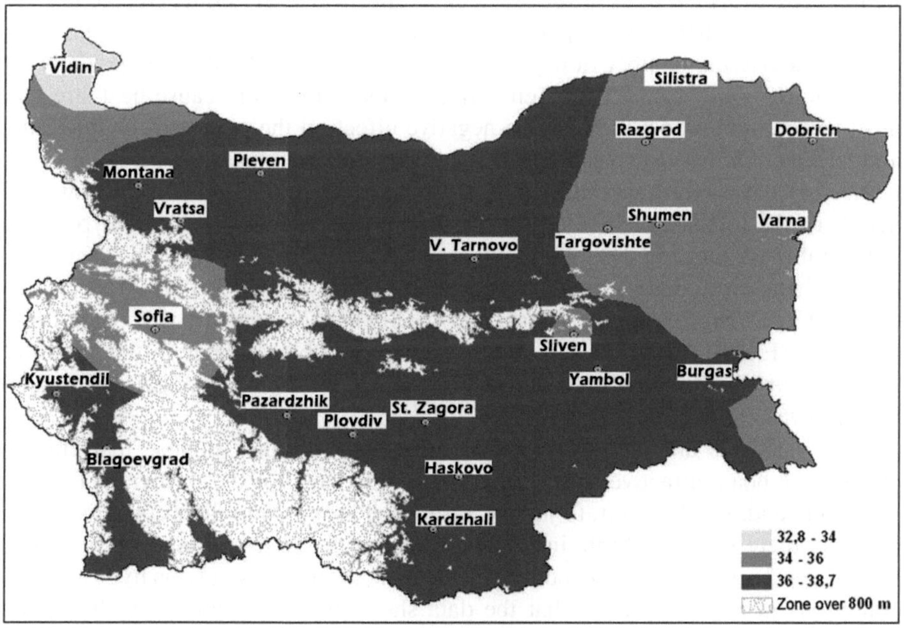

Fig. 3 Distribution of the average maximum air temperatures (°C) for the period 1996–2015

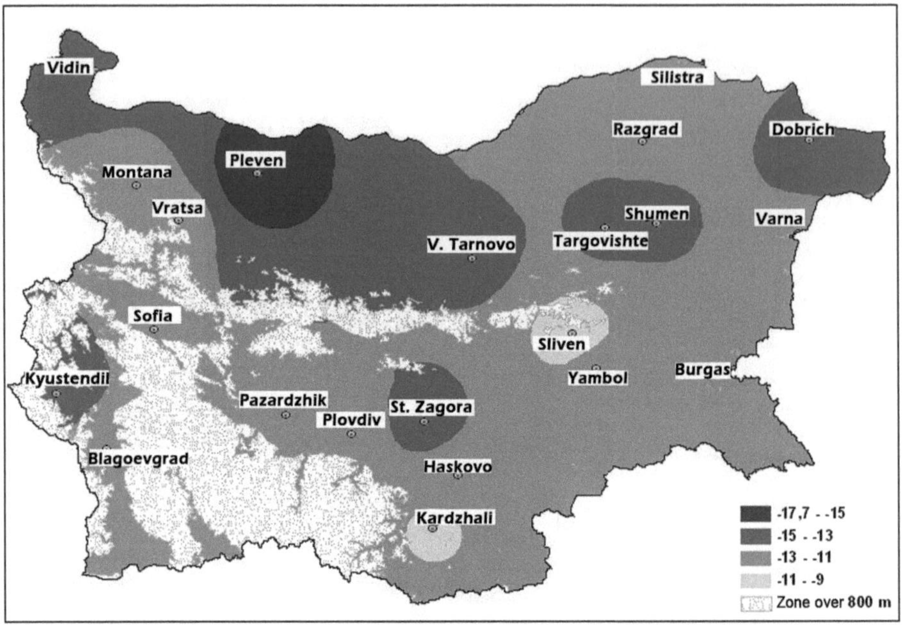

Fig. 4 Distribution of average minimum air temperatures (°C) for the period 1996–2015

years of the twentieth century is 73.7 °C. This proves climate change of continental type to which high amplitudes of temperatures are characterized.

In agriculture it is very important to know not only the length of the growing season but also the number of days of the vegetation period with average daily air temperature higher than 25 °C. Temperatures above this limit cause heat stress in most agricultural crops; they have a negative effect on the processes of flowering and fertilization in most spring crops. Figure 5 shows the spatial distribution of the number of days with temperatures of 25 °C in 1961–1990, and Fig. 6 shows the details of the forecast of increase in number in 2021–2050. The number of those unfavorable for farming days with temperatures above 25 °C will triple. As a result of this, change would mean annual temperature for the period 1961–1990 is between 11 and 13 °C (Fig. 7) in rural areas, and in 2050–2070, forecast will reach up to 13–16 °C (Fig. 8). Under these conditions there will be a change in the length of growing season, which will reach 260–300 days in 2030 (Fig. 9), against 240–260 days in 1961–1990 (Fig. 10) [44–47].

Many analysts are skeptical back in time, and now they oppose that climate change is a hoax attractive to scientists and politicians: for the former, to frighten the public and, for the latter, to justify unfulfilled promises to their constituents. In practice, there is an infallible indicator of natural processes and are phenological events, i.e. the pace of development of plants, which depends solely on hydrothermal conditions in nature. Here's what the data show for our country – in the period 1961–1975, early development of crops has started around the 12th of April, and

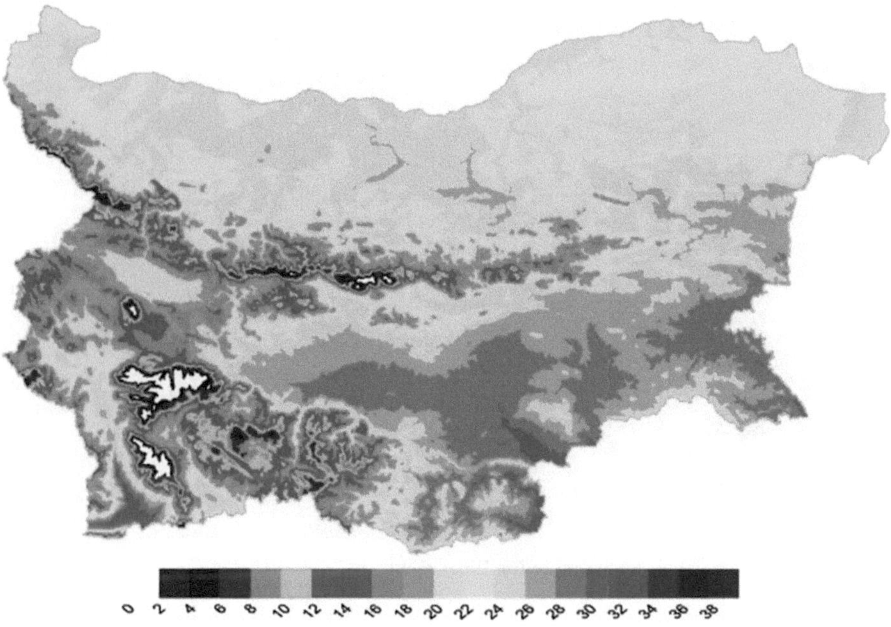

Fig. 5 Number of summer days with maximum temperature >25 °C, 1961–1990

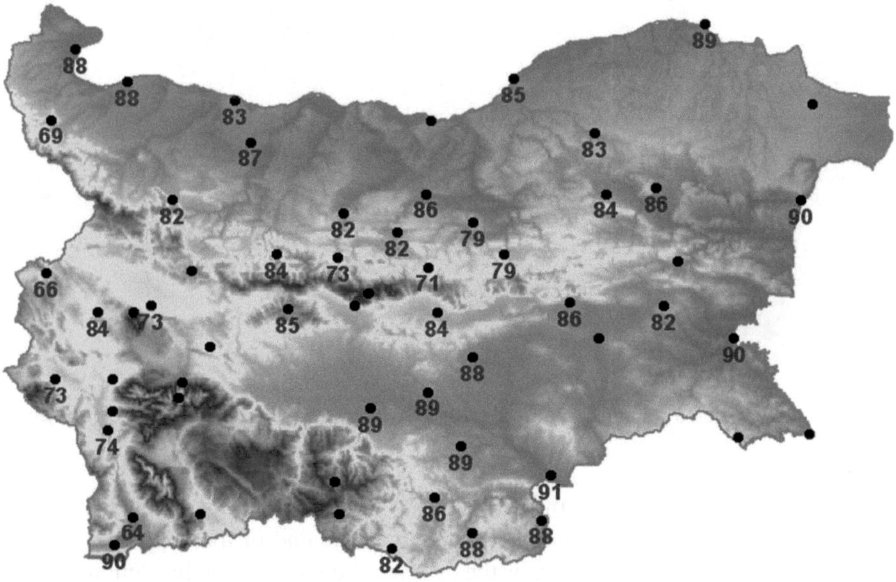

Fig. 6 Number of summer days with maximum temperature >25 °C, 2021–2050, the regional model ALADIN

Fig. 7 Average annual air temperature for 1961–1990

Fig. 8 Average annual air temperature for 2050–2070 under pessimistic scenario for climate change

during the period 1971–2005, the average date for the country is on the 3rd of April or 9 days earlier. Similar phenomenon occurs at the end of the vegetation period, which in this case is 7 days later, or the average growing season has lengthened by 16 days. This fact is certainly evident in the increasing average daily temperatures in 1971–2005 compared to 1961–1975 – Fig. 11. We could mention another

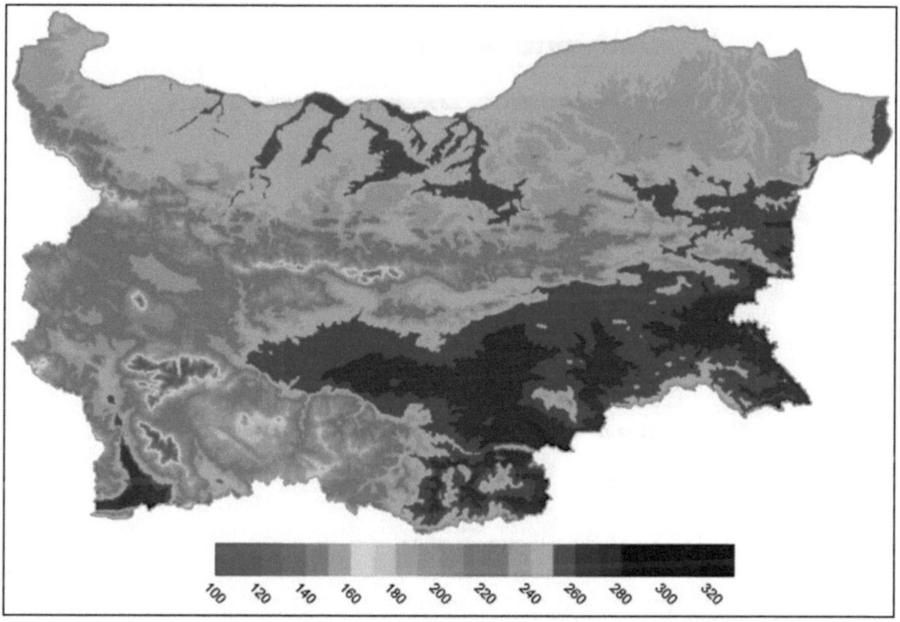

Fig. 9 Expected duration (in days) of real vegetative period (RVP >10 °C), 2030

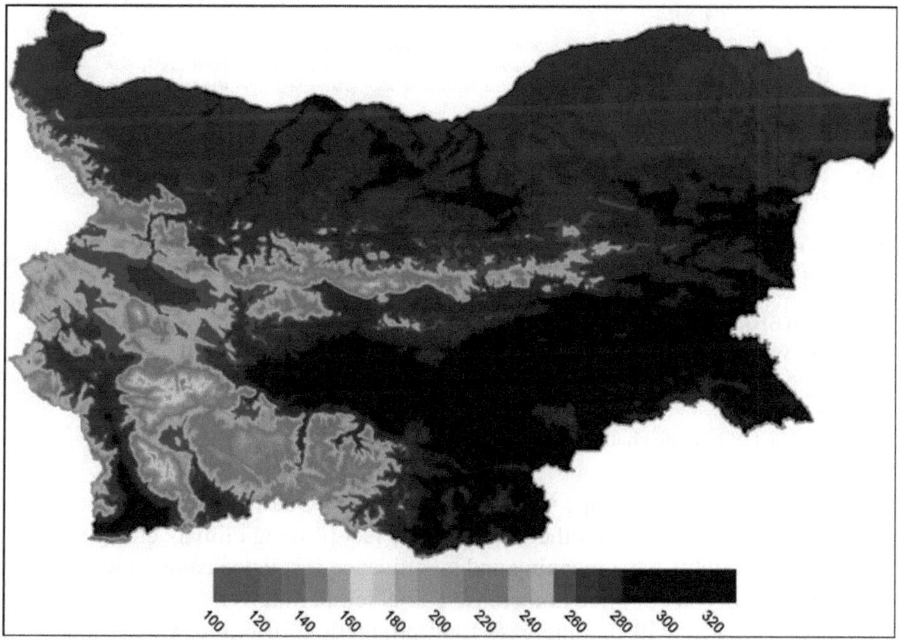

Fig. 10 Duration (in days) of real vegetative period (RVP >10 °C), 1961–1990

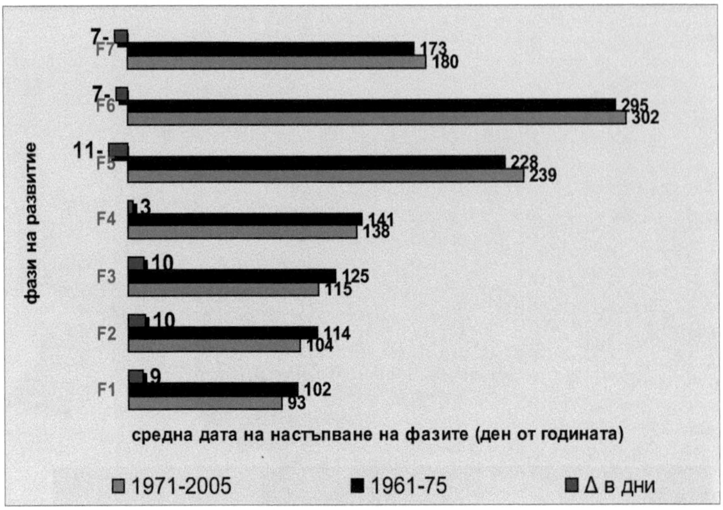

Fig. 11 Phenological stages as indicator for warming

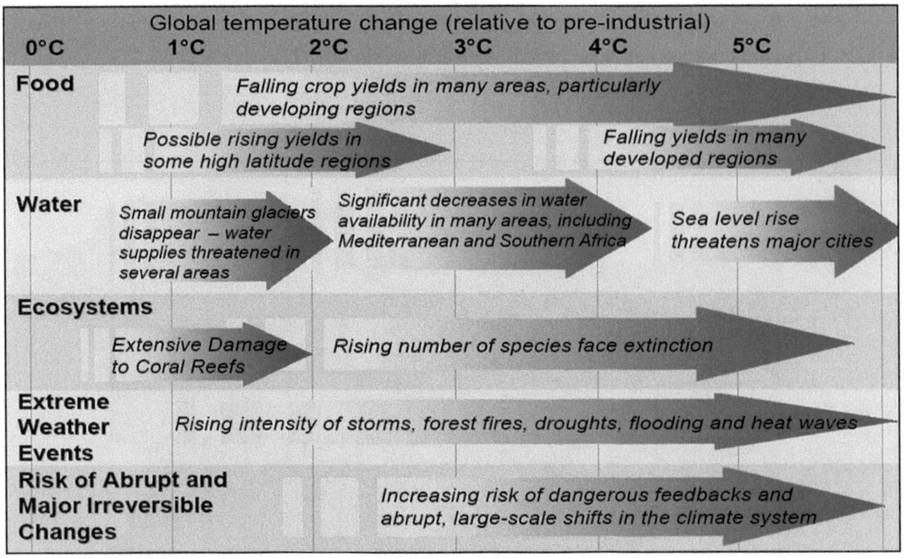

Fig. 12 Impact of climate change

fact – 30 years ago wine grape varieties ripen around October 20. Today they ripen in late September – this is another indisputable fact proving climate change.

All these thoughts have meaning and significance if science cannot answer what would happen if you continue to increase global temperatures by one, two, three, or more degrees. How will this increase food production, drinking water supply, the behavior of ecosystems, and not least the intensity and frequency of extreme weather

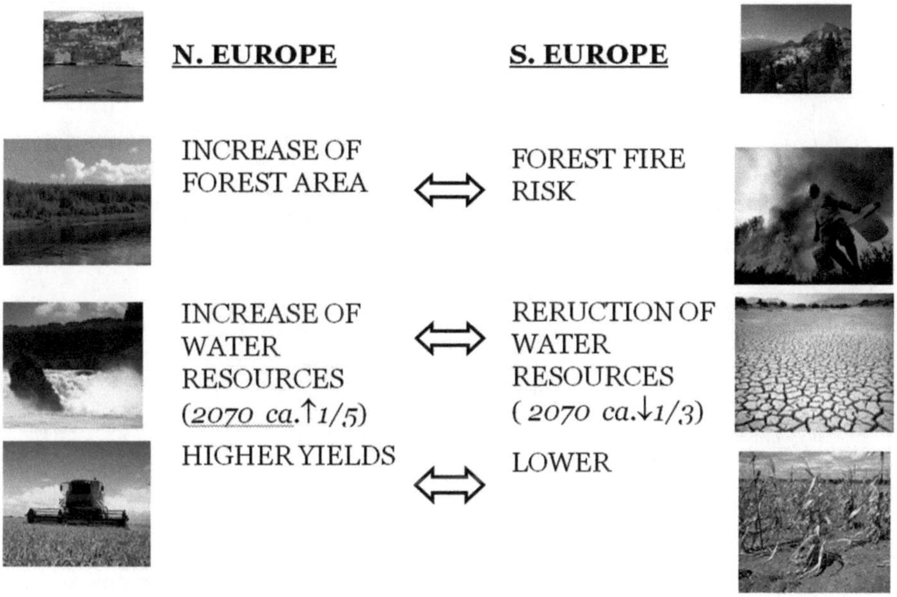

Fig. 13 Differences between Northern and Southern Europe as a result of climate change

events causing calamity (Fig. 12)?. All these changes will cause the change of some sectors in European countries in an irreversible way – Fig. 13.

Moisturizing Conditions

Another very important parameter for characterization of agroclimatic conditions is the rainfall sums. During the last 50 years, the sum of rainfalls in agricultural area in Bulgaria was between 550 and 845 mm, and their spatial distribution is shown on Fig. 14. Spatial distribution of annual sums and deviations in comparison with 1961–1990 is presented on Fig. 15. The minimum of rainfalls is observed in 1985, 1992, 1993, 2000, and 2007 and maximum quantity in the years 1979, 1980, 1987, 1991, 1995, 2002, 2005, and 2012 [34, 37, 41, 43]. **Analysis of spatial distribution of rainfall deviation for the current 50-year period on the territory of the country shows an irregular field of rainfalls. For assessment of moisturizing conditions of agroclimatic resources for vegetative period, it is enough to calculate rainfall sums for these specific time intervals limited by crossing of temperatures through 5 and 10 °C – Figs. 16 and 17** [20–32].

The process of climate change affects all parts of the planet from the poles to the equator. Figure 18 shows the countries according to their industrial and agrarian development and the degree of vulnerability to climate change.

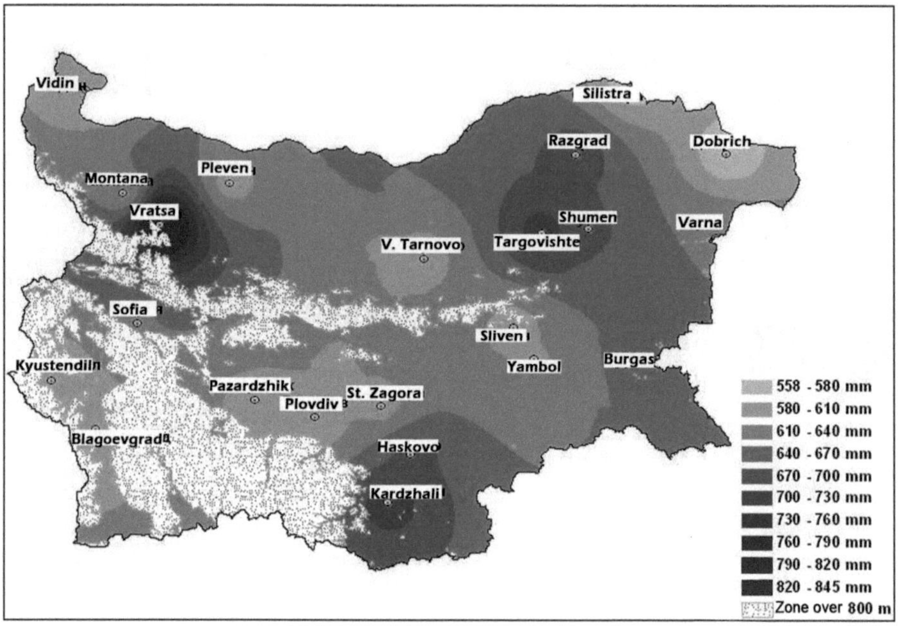

Fig. 14 Climatic values of annual rainfall for the period 1976–2015

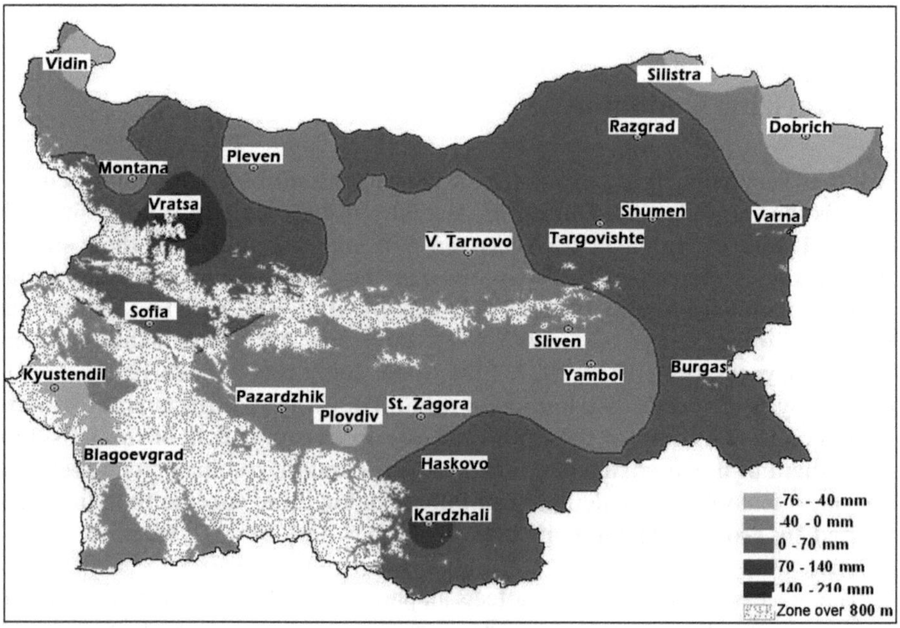

Fig. 15 Deviation of rainfall during the period 1976–2015 to the period 1961–1990

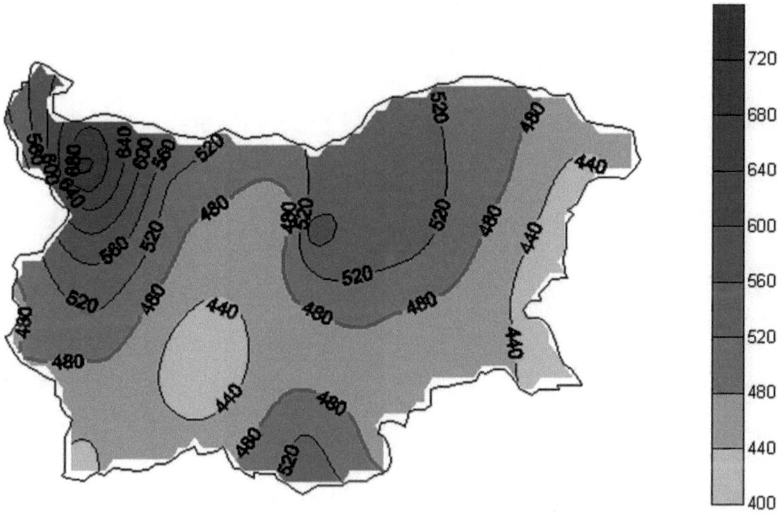

Fig. 16 Spatial distribution of rainfalls during the period with average daily air temperature above 5 °C for 1971–2010 period, thick line correspond to average sum of rains for the country

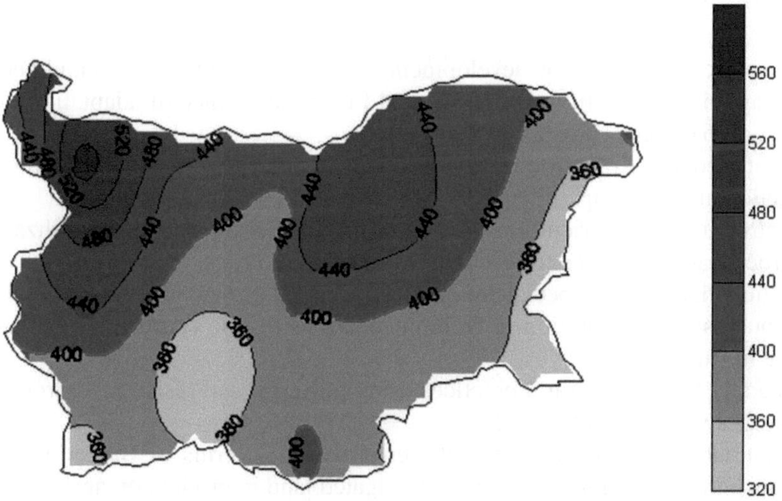

Fig. 17 Spatial distribution of rainfalls during the period with average daily air temperature above 10 °C for 1971–2010 period, thick line correspond to average sum of rains for the country

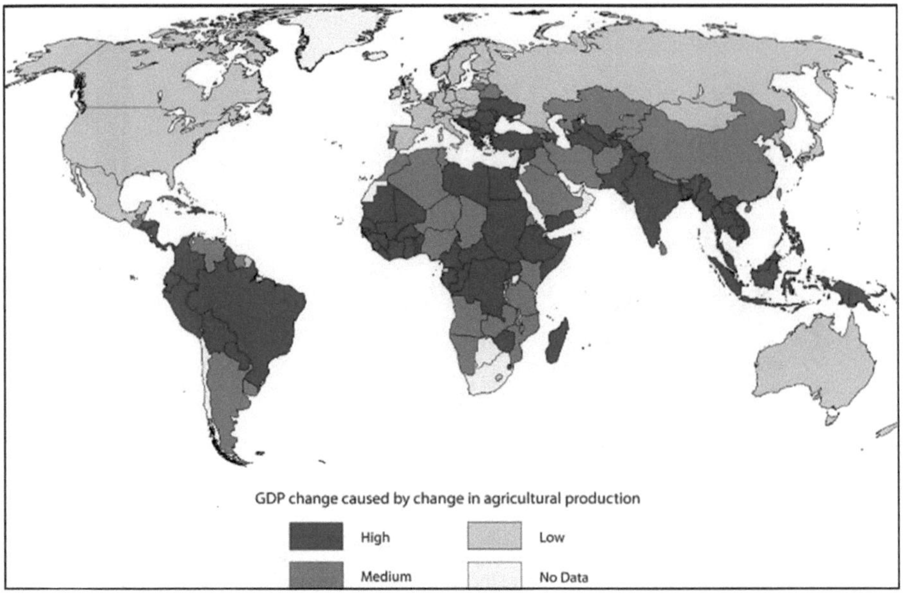

Fig. 18 Agro-economy vulnerability from climate change

Adaptation

- The basic strategy for the development of agriculture in Bulgaria should be complying with results from Bulgarian and European science of adaptation of agriculture to climate change.
- A detailed study of agroclimatic resources in order to improve agricultural production and its competitiveness.
- Restructuring of agriculture in areas with a focus on maximum utilization of agroclimatic resources for growing basic types of crops.
- Regionalization and specialization of agricultural production under agroclimatic resources and preparation of recommendations for the restructuring of agriculture in Bulgaria.
- To apply best agricultural practices using powerful and resistant to drought and diseases varieties and hybrids.
- Sustainable agriculture with high performance in terms of Bulgaria can only grow under irrigation: increasing of irrigated land in production agriculture.

Conclusions

- The agroclimatic conditions in Bulgaria are really various due to the kind of the relief and the climate peculiarity of the country.

- In the period 1966–2015, average annual temperatures have increased themselves with 2.5 °C in comparison with the beginning of meteorological observations and measurements with 0.7 °C in comparison with the 30 years characterizing climate in the twentieth century (1961–1990).
- The rains as well barely change as their quantity has decreased by 150 mm in comparison with the beginning of measurements and by 40–60 mm reasonably in comparison with the previous 30-year period.
- In connection with the showed changes to the climate, huge areas with conditions that are adverse for intensive farming individualize themselves. According to the terminology of JRC, it is called low favorable area (LFA). For the climate change, mitigation of consequences must invest additionally for the farming to these regions.
- For the conditions in Bulgaria, it is necessary to take urgent measures for improvement of soil moisture conditions; that is why high-tech solutions for the irrigation in major part of the vegetative season are needed.
- From this point of view, it is recommended to bear in mind the requirements in growing crops in best conditions.

References

1. Olmstead AL, Rhode PW (2008) Creating abundance: biological innovation and American Agricultural Development. Cambridge University Press, New York
2. Alexandrov V, Dubuisson B, Moisselin J-M, Koleva E (2006) A case study on utilization of precipitation indices in Bulgaria. In: Proceedings of the international conference on water observation and information system for decision support (BALWOIS), Ohrid, Macedonia, 23–26 May 2006, (CD), 18 pp
3. Bachvarova S, Dzhoreva M, Hristozova J, Senyov A (2008) Climatic anomalies and their impact on agricultural production in Bulgaria in 2007 – materials from the work of the National Commission for independent monitoring of the impact of adverse weather events agriculture. MAF, NCAS; NKNM, Sofia, 167 pp. ISBN-978-954-8045-10-0. (in Bulgarian)
4. Croitoru A-E, Chiotoroiu B-C, Todorova V, Torica V (2013) Changes in precipitation extremes on the Black sea Western coast. Glob Planet Chang 102:10–19
5. Rosenzweig C, Hillel D (1998) Climate change and the global harvest: potential impacts of the greenhouse effect on agriculture. Oxford University Press, Oxford
6. De Wit A, van Diepen K, Kroes J, Eitzinger J, Kazandjiev V, Tulios L (2001) Application of remote sensing data as inputs for the SWAP 2.0 model – paper on the COST action 718 meteorological applications for Agriculture, Budapest, p 11
7. Eitzinger J, Thaler S, Orlandini S, Nejedlik P, Kazandjiev V, Vucetic V, Sivertsen TH, Mihailovic DT, Lalic B, Tsiros E, Dalezios NR, Susnik A, Kersebaum CKC, Holden NM, Matthews R (2008) Agroclimatic indices and simulation models. In: Survey of Agrometeorological Practices and Applications in Europe, Regarding Climate Change Impacts, ESF, COST 734, 15–114 p
8. Georgieva V, Kazandjiev V (2013) Climate change and agrometeorological conditions for growing winter crops in Bulgaria. Sci Pares Ser A Agron LVI:459–467
9. Georgieva V, Kazandjiev V, Degorski M, Blazejezyk K, Kucheik M, Degorska B (2015) Investigation on soil moisture reserves and meteorological conditions in relation to basic soil types in Bulgaria. Europa XXI 29:43–58

10. Gocheva A.,K. Malcheva, 2010. Droughtly spells on the territory of Bulgaria. BGMH, V-15, 54–63
11. Gospodinov I, Kozinarova G, Stoicheva A (2010) Very local heavy rain in the Bulgaria Black sea coast in late summer and early autumn. In: Proceedings of the conference on water observation and information system for decision support (BALWOIS), 25–29 May 2010, Ohrid, Macedonia (CD version)
12. Gospodinov I, Dimitrova P, Stoyanova S (2006) Flood event in Bulgaria in August 2005 (the Ichtiman Cyclone). In: proceedings of the conference on water observation and information system for decision support (BALWOIS), 23–26 May 2006, Ohrid, Macedonia (CD version)
13. Hristov H, Latinov L (2006). Three environments in the summer of 2005. Led to flooding in Bulgaria. First national scientific and practical conference on emergency management and civil protection. Sofia – Bulgarian Academy of Sciences 10.11. 2005. Center for Research on National Security and Defence – BAS, 134–141 (in Bulgarian)
14. Nardi JB (2007) Life in the soil: a guide for naturalists and gardeners. University of Chicago Press
15. Lowenfels J (2006) Teaming with microbes: a gardener's guide to the soil food web. Timber Press
16. Kazandjiev B. 2008. Climate change, agro-climatic resources and defining disadvantaged areas in Bulgaria, "The future development of agriculture in Bulgaria," Sirius, Veliko Tarnovo, pp. 139–154 (in Bulgarian)
17. Kazandjiev C (2008) Disadvantaged areas in Bulgaria, determined by climatic indicators and agro-climatic resources during the period 1971–2000, the Expert report for MAF, 50 p (in Bulgarian)
18. Kazandjiev C, Peev B (2005) Prerequisites for disaster by natural weather phenomena and processes, reports first scientific-practical conference on emergency management and civil protection, Sofia, Bulgarian Academy of Sciences 10.11.2005, pp 186–193 (in Bulgarian)
19. Kazandjiev V (2008). Agroclimatic resources and definition of less favored areas at the beginning of XXI century in Bulgaria, Conference "Global Environmental Change – Challenges to Science and Society in Southwestern Europe" (CD version)
20. Kazandjiev V, Roumenina E, Georgieva V, Dimitrov P, Zhelev G (2014) Biometric state of winter wheat agroecosystem determined through ground based measurements and satellite images with different resolution. Balk Ecol 17(2):123–143
21. Kazandjiev V, Degorski M, Blazejezyk K, Georgieva V (2015) Agrometeorological conditions in Bulgaria and agricultural adaptations. Wuropa XXI 29:23–41
22. Kazandjiev V, Moteva M, Georgieva V (2010) Climate change, agroclimatic resources and agroclimatic zoning of agriculture in Bulgaria. Agric Eng 3:109–116
23. Kazandjiev V, Georgieva V (2014) Different indices and characterizing of agrometeorological drought conditions. In: Proceeding 14th international multidisciplinary scientific geoconferences, Sofia, pp 75–83
24. Kazandjiev V, Georgieva V, Moteva M (2009). Climate change, agroclimatic resources and agroclimatic zoning of agriculture in Bulgaria, 2009. 9th EMS/ECAM conference, Toulouse, France http://www.emetsoc.org/fileadmin/ems/dokumente/annual_meetings/2009/AW12_EMS2009-543.pdf
25. Kazandjiev V (2004) Agrometeorological aspects of drought in Bulgaria during last decade (1994–2003). In: Proceedings of the EWRA symposium on "water resources management-risks and challenges for the 21st century", Izmir, Turkey 2–4 September, 2004, pp 331–342
26. Knox JW, Hurford A, Hargreaves L, Wall E (2012) Climate change risk assessment for the agriculture sector in UK climate change and risk (Assessment Defra Project Code GA0204), 252 p
27. Koleva E, Alexandrov V (2008) Drought in the Bulgarian low regions during the 20th century. TAC 92:113–120
28. Sivakumar MVK, Motha RP, Das HP (eds) (2005) Natural disasters and extreme events in agriculture: impacts and mitigation. Springer

29. Sivakumar MVK, Hansen J (2007) Climate prediction and agriculture. Springer, 306 pp
30. Doering OC, Randolph JC Southworth J et al (eds) (2002). Effects of climate change and variability on agricultural production systems. Springer
31. Pringle P (2005) Food, Inc.: Mendel to Monsanto – the promises and perils of the biotech harvest. Simon & Schuster
32. Raev I, Knight G, Staneva M (2003) Drought in Bulgaria: a modern analogue of climate change. Ed. BAS, 284 p (in Bulgarian)
33. Lal R, Uphoff N, Stewart B, Hansen DO (eds) (2005) Climate change and global food security. CRC
34. Few R, Matthies F (eds) (2006) Flood hazards and health: responding to present and future risks. Earthscan Publications, London
35. Roumenina E, Kazandjiev V, Dimitrov P, Filchev L, VassiBGN V, JeBGN G, Georgieva V, Lukarski H (2013) Validation of LAI and assessment of winter wheat status using spectral data and vegetation indices from SPOT VEGETATION and simulated PROBA-V. Int J Remote Sens 34(8):2888–2904
36. Pritchard S, Amthor J (2005). Crops and environmental changes. Haworth Press (US)
37. Simeonov P, Boev P, Petrov R, Sirakov D, Andreev V (1990) Problems of fighting hailstorms. Univ. Ed. "Kliment Ohridski", 314 p (in Bulgarian)
38. Simeonov P, Georgiev CG (2001) A case study of tornado-producing storm south of Rhodopes mountain in the East Mediterranean. Atmos Res 57:187–199
39. Sivakumar MVK, Motha RP, Das HP (2005) Natural disasters and extreme events in agriculture. Springer, 367 pp
40. Slavov N (1998). Relationship between climate change and desertification. Problems of land degradation and combating desertification. UN str.42-48. (in Bulgarian)
41. Slavov N, Raev I (2001) Forest climatic assessment on the duration of the potential vegetation period in Bulgaria. Silva Balkanica 1:35–46
42. Slavov N, Alexandrov V (1993) Drought in Bulgaria. Historical perspective and 1993 update. Drought Netw News 5(2):12–15
43. Stoicheva A, Latinov L (2006) Weather conditions in the winter of 2005. Causing floods and blizzards in some areas of Bulgaria. First National Scientific and Practical Conference on Emergency Management and Civil Protection. Sofia – Bulgarian Academy of Sciences 10.11. 2005. Center for Research on National Security and Defence – BAS, 211–219 (in Bulgarian)
44. Wittwer SH (1995) Food, climate, and carbon dioxide. CRC
45. Shiva V (2000) Stolen harvest: the hijacking of the global food supply. South End Press
46. Cline WR (2007) Global warming and agriculture: impact estimates by country. Peterson Institute (US)
47. Tuba Z (ed) (2005) Ecological responses and adaptations of crops to rising atmospheric carbon dioxide. Food Products Press, New York

Climate Change and Water as a Resource Will Cause Serious Security Implications Around the Globe

Kosta Delev

Abstract Climate change is not a new or surprising event, but it has a significant impact on human civilization. Our future depends on our ability to predict and mitigate the impact and threats created by climate change. The effects will influence all of us. Climate change is a threat multiplier and puts every nation at risk. The aim of this paper is to define the security threats caused by climate change and poor water resource management. The biggest threat is water scarcity. Reduced water resources will create a tremendous impact on all countries, and their economies and societies will react to the stress. It is the responsibility of each government to avoid the conflict.

Keywords Climate change • Water resource management • Balance of power • Security threats • Threat multiplier • Water wars

Introduction

Climate change is not a new or surprising event, but it has a significant impact on human civilization. The future of the people depends on the impact and threats created by climate change which is transboundary and influences the entire world.

When this research began, the author looked for sources which would help to define the root of the problem or to find something similar in the past. It was surprising how often this event had happened. The global climate is constantly changing for a variety of reasons: shifting of tectonic plates, changes in the Earth's axis shift, changes in the Earth's elliptical orbit around the Sun, volcano eruption, and many other random or cyclical events.[1]

[1] Government of Canada, "Canada's Action on Climate Change, Causes of Climate Change," last modified 10 December 2013, accessed 28 September 2014, http://www.climatechange.gc.ca/default.asp?lang=en&n=65CD73F4-1

K. Delev (✉)
NATO Force Integration Unit-Bulgaria, Sofia, Bulgaria
e-mail: Kosta.DELEV@bgr.nfiu.nato.int; kostadinkov@gmail.com

O. Nikolov, S. Veeravalli (eds.), *Implications of Climate Change and Disasters on Military Activities*, NATO Science for Peace and Security Series C: Environmental Security, DOI 10.1007/978-94-024-1071-6_22

This paper will focus on providing a good understanding of the problem, in order to define what the threats from the climate change are and how they will influence and cause serious security implications around the globe.

Climate Change as a Threat Multiplier

Climate change influences the globe and creates a lot of instability and threats. Climate change is not a direct threat but it is a threat multiplier.

Climate change poses another significant challenge for the United States and the world at large. As greenhouse gas emissions increase, sea levels are rising, average global temperatures are increasing, and severe weather patterns are accelerating. These changes, coupled with other global dynamics, including growing, urbanizing, more affluent populations, and substantial economic growth in India, China, Brazil, and other nations, will devastate homes, land, and infrastructure. Climate change may exacerbate water scarcity and lead to sharp increase in food costs. The pressure caused by climate change will influence resource competition while placing additional burdens on economies, societies, and governance institutions around the world. These effects are threat multipliers that will aggravate stressors abroad such as poverty, environmental degradation, political instability, and social tensions – conditions that can enable terrorist activity and other forms of violence.[2]

This quotation from US *Quadrennial Defense Review* 2014 gives a concise picture of how climate change acts as a threat multiplier and influences the globe. This is an example of how the military see the threats that can create conflicts because of the climate change.

World leaders met to tackle climate change during the General Assembly of the United Nations (UN) on 16–30 September 2014 in New York. During his speech, the UN Secretary-General Ban Ki-moon underscored the importance of climate change as the defining issue of our age and called for leadership to cut greenhouse gas emissions.[3] In his speech on 15 January 2015, the President of the United States of America Barack Obama said, "No challenge poses a greater threat to future generations than climate change."[4]

Climate models show significant agreement for all emission scenarios in warming (magnitude and rate) all over Europe, with strongest warming projected in Southern

[2] Department of Defense, *Quadrennial Defense Review 2014* (Washington, DC: Government Printing Office, March 4, 2014), 8.

[3] UN News Center, "Ban Ki-moon's Speeches: A Quick Guide to Statements on Key Issues on UN Agenda," UN News Service, last modified 16 September 2014, accessed 28 September 2014, http://www.un.org/apps/news/infocus/sgspeeches/

[4] Madison Park, "Obama: No Greater Threat to Future Than Climate Change," *CNN*, last modified 21 January 2015, accessed 4 May 2015, http://www.cnn.com/2015/01/21/us/climate-change-us-obama/

Europe in summer and in Northern Europe in winter. Even under an average global temperature increase limited to 2 °C compared to preindustrial times, the climate of Europe is simulated to depart significantly in the next decades from today's climate.[5]

It is obvious from the quotation that climate change will have a significant impact on the European continent. In Southern Europe, the impact of climate change has already started. The extreme weather conditions in the region – floods, droughts, heat waves, and intense storms – have increased dramatically and have turned into threats for the population.

On 06 June 2016, the NATO Secretary-General Jens Stoltenberg participated in an online discussion at the POLITICO Playbook Cocktails at Residence Palace in Brussels. During the interview he said:

> JENS STOLTENBERG: *First of all I think it's very important to underline what you just said, and that is that climate change is also a security threat because it can really change also the conditions for where people can live, it can create new migrant and refugee crises and scarce resources, water can fuel new conflicts. So climate change is also about preventing conflicts, and creating more stability and prosperity which is good for peace and stability.*[6]

This happened prior to the Warsaw Summit, where NATO agreed that climate change is a threat multiplier and it has a direct impact on military operations.

The threats created by climate change will influence the entire world. Predictions from various sources show that by the end of this century, the world population will increase to more than ten billion people, and the world temperature will increase by at least 2 °C. As a result, ocean levels will rise by as much as 1 m and freshwater resources will decrease dramatically. According to the Intergovernmental Panel on Climate Change (IPCC) *IPCC – Climate Change 2014 Impacts, Adaptation, and Vulnerability* (hereafter referred to as IPCC 2014), "For each degree of global warming, approximately 7% of the global population is projected to be exposed to a decrease of renewable water resources of at least 20% (multi-model mean)."[7]

For example, the population on the African continent will double, and it will need more food, freshwater, and land. This future environment with more people, higher temperatures, rising ocean levels, and fewer freshwater resources will create new and additional types of problems.

---

[5] Intergovernmental Panel on Climate Change, *IPCC - Climate Change 2014 Impacts, Adaptation, and Vulnerability Part B: Regional Aspects*, edited by Vicente R. Barros and Christopher B. Field (New York: Cambridge University Press, 2014), 1276.

[6] North Atlantic Treaty Organization, Newsroom, Speeches & transcripts, Accessed July 18, 2016: http://www.nato.int/cps/en/natohq/opinions_132050.htm

[7] Ibid., 232.

Water as a Source for Future Conflicts

Because of these dramatic changes, water will become more precious and important. These changes will create tensions among neighboring countries because of poor water management or lack of it.

In the book *The Global Water Crisis: Addressing an Urgent Security Issue*, the author poses the question, "Will the Next Wars Be Fought over Water?" Bob Sandford provides a lot of information and conclusions about the past and future conflicts based on water:

Some 181 conflicts over water are reported to have occurred between 3000 BC, and the end of 2007. Some 146 of these conflicts took place in the 5000 years between 3000 B.C. and the year 2000. The remaining 59 conflicts therefore occurred in this century. During that same brief decade, new forms of actual and potential conflicts over water emerged. These include homegrown terrorist threats to water infrastructure in Afghanistan and Iraq and a foreign terrorist threat issued by Al-Qaida in 2003 against domestic water supply systems in the United States.[8]

A great majority of the world's conflicts has started or will start for resources and territories. We should take into consideration that territory issues are linked with the resources that are on a particular territory.

A careful assessment of this history of conflict reveals that while water systems have been used as weapons and targets during war, water resources in themselves have rarely been the sole source of violent conflict or war. This has led water scholars to maintain that – since the 1940s, at least – water is more than twice as likely to be a source of international cooperation as of conflict. But as Peter Gleick points out, the fact that there has been widespread international cooperation over water should not allow policy-makers to underestimate the complexity of the relationship between water and national security.[9]

Two points that can be extracted from the information: first point is that water can be used as a weapon and it can be a source for violent conflicts; second point is that water is important for any national security. Next, Sandford presents an idea of how these "water wars" can be avoided or diminished.

There are a number of factors that reduce the risk of traditional water wars, such as the presence of new transnational institutions like the United Nations, more effective international laws, the emergence of the International Court of Justice, more comprehensively crafted treaties, new water conservation measures and technologies, and better dispute resolution mechanisms. This hope, however, is founded upon the anticipated stability, or rather stationary, of both demand for and reliable

[8] Gro Harlen Brundtland, *The Global Water Crisis: Addressing an Urgent Security Issue*, edited by Harriet Bigas, Tim Morris, Bob Sandford, and Zafar Adeel (Hamilton, Canada: UNU-INWEH, 2012), 12.

[9] Gro Harlen Brundtland, *The Global Water Crisis: Addressing an Urgent Security Issue*, edited by Harriet Bigas, Tim Morris, Bob Sandford, and Zafar Adeel (Hamilton, Canada: UNU-INWEH, 2012), 12.

availability of global water supplies. Unfortunately, our global hydrological situation is changing rapidly and may soon no longer resemble anything that has existed on Earth before, at least in human memory.[10]

The importance of this block is to predict whether conflicts, based on water, may occur and how such conflicts may be reduced or avoided. He suggests a few ways to do that: to use international institutions, to sign agreements, to establish better dispute mechanisms, and, finally, to use new technologies to reduce the loss and increase the effectiveness of water use.

It is determined that water security can be a source of future conflicts: more precisely, a shortage of water or floods that may cause state failure and increase regional tensions. Furthermore, water can be used as a weapon or as a source of violent conflicts. Water directly influences population, economy, health system, agriculture, energy, peace, and political stability.

Achieving peaceful solutions must come through the use of international institutions, international agreements, establishing better dispute mechanisms, and new technologies to reduce the loss and increase the effectiveness of water use. This means that with agreements and developed water resource management, conflicts can be avoided or resolved.

Balance of Power

The analysis on the effects of climate change and water management and agreements shows direct links to balance of power among nations. The balance of power has a direct link with the balance of sharing and using resources. Water is a resource that one may share or may be forced to share. This is the most distinguished difference: to share or to be forced to share. People can use appropriate and fair resource management to support this balance, or the situation can force them to do that. This is the area where one can link and search for security threats to the system.

The perfect situation for a balance has not happened for a long period of time. A lot of factors can influence this stable situation and make it unstable. It can be population growth or decline, discovery of new resources, decline of existing resources, environmental changes, social system changes, ethnical homogeneity, and many others.

A change in environment is the foundation of all other changes. Climate change, as a threat multiplier, indirectly changes the balance of resources which leads to changes in environment. The crucial resource for the system is water.

[10] Ibid.

Conclusion

Climate change, as a threat multiplier, will increase the level of threats all over the world. If action is not taken to adapt and mitigate these threats, they will increase every single year and will start to threaten our existence. Simultaneously, the cost to rebuild or fix the damaged infrastructure will increase; alas, the most dangerous will be the human cost. Society will pay this price if political leaders around the globe are not ready to implement new policies for adaptation and mitigation. Political, economic, educational, energy, health, and security systems should start to prepare for and build up new capability and capacity for the future.

Climate is a global issue and the changes are global, too; this means, climate change will affect every place in the world. No matter in which continent your country is or how green your society lives, the change will affect you. The only way to decrease the damages and the causalities will be to start to prepare the society, the country, or the region to adapt and change for the new reality.

If the problem is identified, but its roots are not targeted and healed, it will develop into a more serious one. Finally, the price to solve it will be high, and it will increase proportionally to the threats.

The political will has to change and adapt to this new environment. Political leaders have to lead their societies, inform them of the threats, and unite the nations and regions in their effort to deal with that. There is no simple answer to complex problems: to find their roots, we have to study the problems in close details.

The key for success is in the agreements among countries, the will to share resources equally, and the readiness to work together. To unite countries with different cultures and languages is a challenge, but it is not impossible. The EU is a good example of that.

To summarize: climate change, as a threat multiplier, will change the global environment and it will be followed by a resource crisis. The biggest threat is water scarcity. Reduced water resources will have a tremendous impact on our countries: their economies and societies will react to the stress, and it is a responsibility of each government to avoid the conflict. The only peaceful solution is to work together for the good of all nations.

Summary Findings, Recommendations and Conclusions

Orlin Nikolov and Swathi Veeravalli

Abstract Resource managers must consider climate change during their planning process as future landscape conditions may vary greatly from conditions currently observed. Small-scale or regional susceptibility analyses are often too broad to provide meaningful results at the site-specific scale required by managers, while largescale spatial analyses lack transferability and are often not applicable outside a specific study location. Attendees presented and shared currently available and future capabilities for assessing the implications of climate change for civil-military asset readiness to support disaster forecasting and response. Case studies were demonstrated multi-disciplinary capabilities (e.g., engineering, earth science, social science, policy) that can enhance the deployment of civil-military assets and infrastructure specifically for natural or man-made disaster forecasting and response. This workshop evaluated these current and future capabilities by further defining the problems and challenges that climate change poses to civil-military missions and objectives in the Balkans and over other disaster prone areas throughout Eastern Europe.

Introduction

The Crisis Management and Disaster Response Centre of Excellence (CMDR COE) conducted an advanced research workshop "Climate Change Implications on Military Activities in the Balkans Region" between 05 and 07 July 2016. The event was supported by the NATO Science for Peace and Security (SPS) programme and gathered distinguished subject matter experts from various international organizations and civil-military agencies.

O. Nikolov (✉)
CMDR COE, Sofia, Bulgaria
e-mail: orlin.nikolov@cmdrcoe.org

S. Veeravalli
Geospatial Research Laboratory, US Army Corps of Engineers, Washington, DC, USA

O. Nikolov, S. Veeravalli (eds.), *Implications of Climate Change and Disasters on Military Activities*, NATO Science for Peace and Security Series C: Environmental Security, DOI 10.1007/978-94-024-1071-6_23

Prior to the event, an open survey was developed and uploaded to CMDR COE website in order to assess participant views on climate security challenges within the Balkans. The results revealed concern related to the global security environment and the environmental security in the Balkan region. Further, survey participants were asked to reflect on identifying the five main security challenges over the next 5 years. The following main challenges were highlighted: illegal immigration, economic fragility, terrorism and violent extremism, climate change and natural disasters.

Purpose and Goals

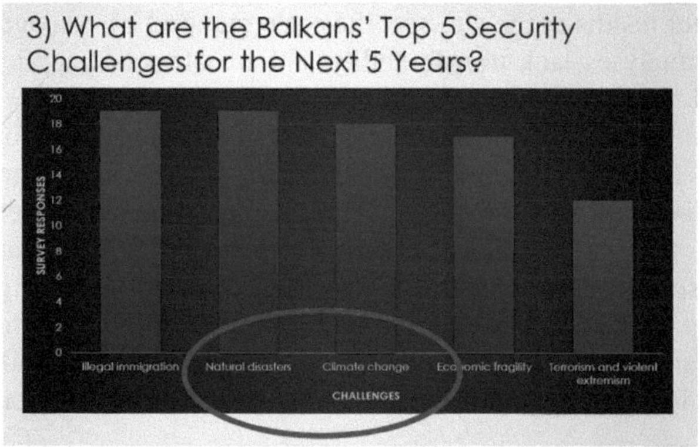

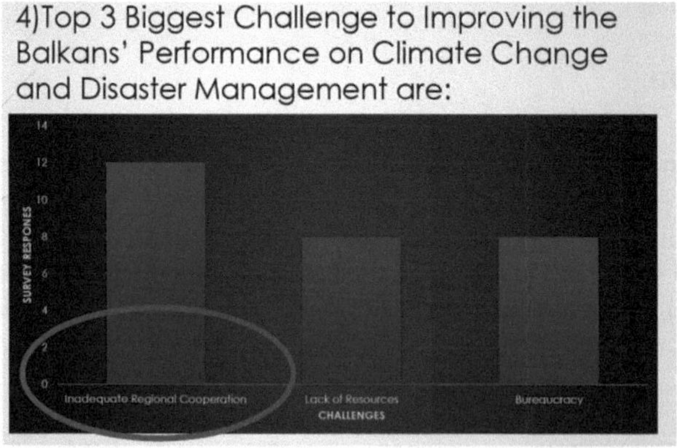

(a) *Purpose*. This 3-day workshop is designed to provide preliminary recommendations on ways to educate and develop capacity among disaster response, security and other professionals from diverse backgrounds whose current and future interests relate to crisis management.
(b) *Goals*. All workshop participants are to understand the threat of climate change and of disasters to environmental security using a multidisciplinary approach to

improving regional security cooperation and to address the complex issues of climate change and disasters on military activities.

(c) *Deliverables*. Main products of the event should be as follows:

1. Education and individual training activity common core curriculum which the main purpose is to support increasing climate change implication awareness.
2. Advanced research workshop proceedings encompassing articles related to the topics presented by the main subject matter experts (SMEs) during the event – this document is to be issued by CMDR COE in accordance with NATO Science for Peace and Security programme requirements.

Overall Evaluation

During the workshop, a survey encompassing all workshop participants was conducted in order to support our work, an evaluation form (EF). The main goal of this document was to provide more information on the event's content, speakers, the most beneficial aspects of the workshop, interesting topics, etc. Thirty-five EFs were filled in by workshop participants. EF results were (maximum score of 5) as follows:

1. Workshop content – 4.65
2. Registration process – 4.82
3. Venue – 4.85
4. Food and beverages – 4.88

All filled EFs can be provided to NATO SPS programme in pdf format as a reference document (provided electronically as a scan document).

Audience

The workshop was attended by participants from 17 nations (ALB, BGR, BIH, BRA, GBR, GEO, GRC, HUN, MNE, NLD, POL, ROM, SRB, TUR, UKR, the USA and FYROM). It brought together around 50 (fifty) participants from various organizations including US European Command; US Africa Command; US Army Corps of Engineers; US Army Command and General Staff College; Asia-Pacific Centre for Security Studies; US Agency for International Development; International Alert; Tennessee Emergency Management Agency; Tennessee National Guard; Environmental Law Institute; Data for Resilience; National Institute for Meteorology and Hydrology, Bulgarian Academy of Science; Directorate General Fire Safety and Civil Protection, Ministry of Interior of the Republic of Bulgaria; and Hellenic National Defence General Staff.

Event Programme

Workshop Purpose

This 3-day workshop is designed to provide preliminary recommendations on ways to educate and develop capacity among disaster response, security and other professionals from diverse backgrounds whose current and future interests relate to crisis management.

Workshop Goals

- To understand the threat of climate change and disasters to environmental security using a multidisciplinary approach to improving regional security cooperation
- To address the complex issues of climate change and disasters on military activities

Day 1: The Changing Environmental Security Landscape Due to Climate Change

Objectives

Identify broad issues on climate change and disasters particularly those with the highest importance and relevance to regional security; identify areas of common interests among participants for future areas of collaboration.

Activities

In the opening session, a welcome address was delivered by *Mrs. Zlatina Karova, director of "Science" Directorate, Ministry of Education and Science of the Republic of Bulgaria*, who emphasized on the importance of events focusing on the regional security implications of climate change and providing grounds for broad cooperation and exchange of good practices among professionals in the respective areas. Mrs. Karova added that for the last 20 years, Bulgaria has actively participated in many scientific projects, and there was a well-established understanding of the significance of potential climate change implications on military activities.

The co-chair of the workshop *Colonel Katarina Strbac representing the Serbian Ministry of Defence* underpinned human activity and globalization as the two key tendencies causing concern in the current environment. Furthermore, she stated that climate change is a result of human activity. Colonel Strbac remarked on the changing characteristic of peace and security and on the asymmetric nature of contemporary threats noting the critical importance of adapting military responses to the imperatives of the new security environment. She also underscored the importance of communication and cooperation between the scientific community and the military for undertaking appropriate and effective measures in response to ever increasing in number and intensity emergencies in the Balkan region. Community preparedness and resilience, particularly in the context of urbanization, are key elements of any emergency response plan aiming at mitigating the risks of loss of lives and infrastructure damage. Moreover, regional cooperation in education and prevention would highly contribute to mitigating the knock-on consequences of emergency situations and disasters. Colonel Strbac added that harmonization plans, including policies and standards (actions) at subregional, regional and global levels, are essential for tackling emergencies having widespread transborder repercussions. In this sense, Colonel Strbac stressed that the ARW provides excellent opportunity for stimulating critical thinking and discussion regarding the challenges within the security sector.

Ms. Swathi Veeravalli, US Army Corps of Engineers, placed emphasis on the aim of the workshop to foster dialogue and encouraged knowledge sharing on climate security among all participants. She noted that whilst NATO is concerned about the security implications of climate change, more and education and training are paramount in addressing the new security challenges. Moreover, she outlined that the ultimate goal of the event was to create a multidisciplinary environment, to improve regional cooperation in the Balkan region, to learn from military and civilian experiences and to finally exchange knowledge and best practices between military and civilians to address the security implications from climate change.

Professor Valentin Kazandzhiev, from the National Institute for Meteorology and Hydrology – Bulgarian Academy of Science, delivered a keynote address, which provoked lively discussion. Special attention was drawn to several key questions: What are the consequences of climate change? How to mitigate the effects of climate change? How to adapt sectors of the economy to climate change? And does national security depend from climate change? Professor Kazandzhiev also emphasized on the unsettling studies that expect extreme weather events to double by 2020 and to triple by 2050, whilst crop yield by 2080 will drop by almost −30%.

He highlighted the following key recommendations:

1. Aligning the basic strategy for the development of agriculture in Bulgaria with the results from Bulgarian and European scientific research on the adaptation of agriculture to climate change.
2. Conducting a detailed study of agroclimatic resources in order to improve agricultural production and competitiveness.

3. Restructuring of agriculture in areas with a focus on maximum utilization of agroclimatic resources for growing basic types of crops is vital.
4. Regionalization and specialization of agricultural production under agroclimatic resources and recommendations for restructuring of agriculture in Bulgaria are needed.
5. Achieving and maintaining sustainable agriculture through increased irrigation in production agriculture.

In the question and answer session, Professor Kazandzhiev provided comprehensive explanations to questions concerning mitigating climate change implications reflecting upon topics such as deforestation, irrigation and agriculture lands and also the support provided to governmental institutions.

Dr. Wendell King, US Army Command and General Staff College, stressed that climate change has already happened and the biggest threat for the Balkan region, especially in the southern part, is water scarcity. He further provided an array of climate change examples, inter alia, the global increase of temperature, air current changes, water scarcity, population growth, overutilization of land, etc. He also briefed on the political aspects of climate change in the USA, provided an overview of processes at the strategic level. Dr. King presented the work of two of his master students, Major Kosta Delev and Captain Kuman Gerovski dealing with climate change implications in the Balkan region. Finally, Dr. King established a correlation between environmental security and state failure without, however, pointing to the direction of this correlation. In the lively discussion which ensued, Dr. King accented that our actions today will have direct impact in the long term and that intelligence gathering and sharing is critical in terms of mitigating the negative impact of climate change to security.

Ms. Janani Vivekananda, from International Alert, took a peacebuilding perspective to climate change, discussing it as a cross-cutting complex issue interrelated with all aspects of the security environment. She presented seven compound climate-fragility clusters, stressing that 80 % of the disasters occurred in fragile states, which is indicative of the interconnection between fragility and disasters. She added that climate change effects power dynamics (control over resources) and that the response of the international community to disasters, providing as reference Pakistan, may have unintended consequences and may if not well coordinated and planned further entrench (unequal) power dynamics. As concrete example of climate change impacts on communities at risk and most vulnerable groups within these communities, Ms. Vivekananda pointed to the quality and availability of potable water and to the rises in the sea level. In conclusion, Ms. Vivekananda once again underlined that fragile states are particularly exposed and vulnerable to the climate changes.

Mr. Jeffrey Andrews, from US Africa Command, introduced the mission and role of the organization he represents and the climate change adaptation roadmap. He explained that efforts are devoted to conflict prevention, building resilience and assisting adaptation through civil-military dialogue and collaboration. Mr. Andrews also stressed on the importance of training on climate change, disaster and security, particularly in cooperation with relevant international/regional organizations. Holistic approach to climate change and security and tapping into the capabilities of the military for non-kinetic actions were presented as critical steps in the process of devising adequate approaches to mitigate climate change consequences to security.

Dr. Carl Bruch, International Law Institute, identified environmental security, human security and disaster risk reduction as key security concepts. He further discussed in more details the disaster risk management process and stressed that contingency planning is the key for prevention and mitigation.

Major Stoyan Stoyanov, representing the CMDR COE, outlined the importance of the correct use of common terminology in accordance with broadly recognized definitions of terms with the aim of avoiding misunderstandings. He presented climate change from NATO's point of view as well as the organization's strategic foresight analysis. In addition, the security aspects of melting polar caps, droughts and extreme weather were also highlighted.

During day 1 panels, the presentations of Professor Valentin Kazandzhiev and Dr. Wendell King extremely contributed for increasing climate change awareness and were highly appreciated by the audience.

All panellists were invited to collectively answer questions in a joint Q&A session which was moderated by Ms. Swathi Veeravalli. Ms. Veeravalli also presented a brief overview on how system thinking could be utilized in disaster risk management to ensure more coordination and cooperation in disaster response and preparedness.

Finally, Dr. Imes Chiu, from the Asia-Pacific Centre for Security Studies, wrapped up the panel highlighting the key takeaways of the first day of the seminar: complexity, change, critical thinking, causal loop, common problems, common solutions, coordinate, collaborate, cooperate and communicate.

Day 1 Accomplishments

- Recognize the importance of a human security approach to disaster risk management.
- Recognize challenges with civil-military cooperation, but there are more opportunities if we can do this utilizing DRM cycle.

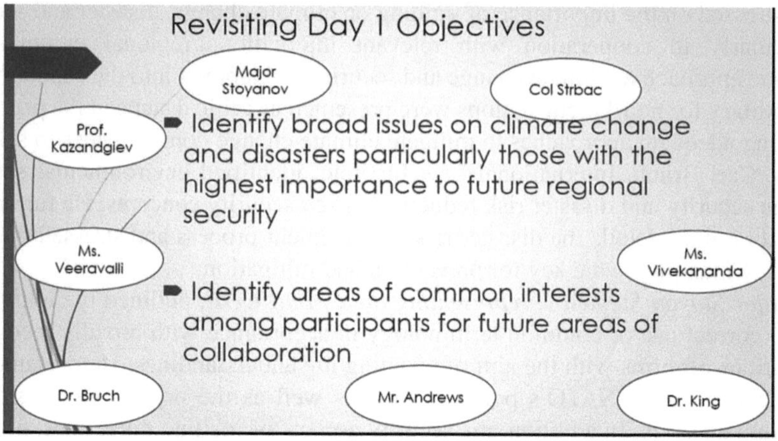

Day 2: Disaster Response (Planning Multiple Stressors Using Whole of Government Approaches)

Objectives

Open exchange of different perspectives on capability gaps and practical analysis of and approaches to disaster management. Understand the importance of surge capacity in complex catastrophes.

Activities

The second day of the workshop further developed the topic of disaster response and provided for experience sharing among participants on managing complex catastrophes/emergencies.

Mr. Georgi Petrov, Directorate General Fire Safety and Civil Protection of the Ministry of Internal Affairs, presented the Bulgaria peer review process in disaster risk management (DRM). This review serves as a tool for facilitating the exchange of good practices, for strengthening mutual learning and common understanding as well as for delivering credible and trusted recommendations. He presented in depth the principles (mutual trust, voluntary basis, value sharing, coherence, commitment and participatory and multi-stakeholder approach), objectives (improved policy-making on national DRM, enhanced policy dialogue in Europe), types (general and thematic), framework and follow-on actions related to the Bulgaria peer review. The key findings presented focused on several key topics:

- Shifting from a response-focused emergency management system to a more holistic disaster management system
- Establishing disaster risk reduction councils to the council of ministers, the regional governors and the mayor levels
- Strengthening the role of the national disaster risk reduction platform through regulating the horizontal and vertical relations at the different levels of governance
- Establishing disaster risk reduction planning at national, regional and local levels
- Optimizing the planning of the effective use of the financial resources
- Promoting and encouraging both public and private sectors to undertake actions to address the disaster risk management activities

Most of the aforementioned recommendations/findings led to respective amendments in the Bulgarian Disaster Protection Act.

Ms. Janani Vivekananda briefed on the resilience system analysis and the use of system approach to examine how a society system could react to disasters. This approach is designed to support the development of strategic vision in various policy areas as it provides a shared view of the risk landscape in a given location, how this risk landscape affects the social system (well-being of the populace) and what are the system's coping mechanism. Moreover, the analysis looks into conflict sensitivity so as to identify root causes and potential triggers that might coincide with climatic shocks, different actors (stakeholders) and power dynamics. Ultimately, the analysis serves to inform policymaking (controlling for uncertainties) and provides basis for the development of (response) scenarios. Ms. Vivekananda also underlined that there are many challenges, particularly in the long term, with respect to collecting accurate and reliable climate change- and conflict-related data. However, she reiterated that such tools and systems provide better understanding of the underlying consequences/factors which may trigger conflict and the aggravating impact of coinciding climatic changes which require timely and targeted response.

Colonel Doug Brantley, Tennessee National Guard, discussed the role of the military in disaster response and pointed out that the military is called upon as a last resort, in case that the civilian capacities for disaster response have been exhausted.

He also pointed out that the military is best suited to support four out of 16 emergency support functions – communications, infrastructure, human services and law enforcement. Specific areas of action presented were the provision of equipment (vehicles, satellites, information systems), route clearance (debris removal capabilities), assistance with shelter and mass care operations and traffic control (security and crime control). Additionally, Colonel Brantley identified five lines of military involvement in case of emergency: critical infrastructure, population protection, mobility, recovery and information operations.

Mr. Luan Qafmolla, from the Albanian Institute of Applied Nuclear Physics, provided a detailed analysis of approaches to building resilience and to mitigate vulnerabilities caused by radioactive sources used in Albania. He also shared the Albanian experience in disaster management in the event of nuclear/radiological emergencies.

Mr. Michael Campbell, US Army Corps of Engineers, provided examples of open-source software and imagery for regional land cover mapping. Mr. Campbell presented applications of open-source geospatial software, combined with freely available space-based digital imagery, for land use/land cover mapping. His thesis was that all phases of the crisis management and disaster response cycle, including planning, prediction, preparedness, response and reconstruction, benefit from up-to-date land cover maps. Mr. Campbell offered two specific tools for acquiring and processing free multispectral earth imagery: Quantum (Q) GIS and Google Earth Engine. His two example land cover products included a large region centred on Belgrade, Serbia, as well as Bulgaria's Black Sea coastline near the city of Varna.

The following Q&A session which ensued saw the panellists respond to a variety of questions concerning the political considerations in disaster/conflict management and the lack of sufficient prevention, the major enablers and challenges for improving cross border cooperation in the Balkan region and the possibility of creating a disaster risk reduction communication platform, the prejudices with regard to the role of the military in disaster management activities and the (Albanian) policies and strategies for radioactive material disposal.

The afternoon session provided for highly interactive participation in the form of a brainstorming on three questions tightly related to the anticipated outcome of the workshop, namely, the development of a course on the implications of climate change on military activities. The questions focused on ways for promoting regional cooperation, target audience, main requirements/justification and indicators to measure against the utility of the course.

Some of the recommendations made detailed the usefulness of a short presentation session in the beginning of course, of regionally focused tabletop exercise and discussions.

Dr. Imes Chiu made a comparison between the response measures undertaken by the Philippines during typhoon Haiyan (2013) and typhoon Hagupit (2014). She further remarked on the necessity to bring science to the people and to ensure multinational multi-stakeholder coordination and cooperation with the aim of saving lives. Connectivity and partnerships, including with the private sector, were also flagged as key elements of disaster management and climate change.

In conclusion of the afternoon session, recommendations were made for the course to have a short presentation session in the beginning of course, tabletop exercise and discussion panel that should be regionally focused.

In day 2, Colonel Doug Brantley and Mr. Michael Campbell delivered briefings that supported in great extend all subsequent workshop activities.

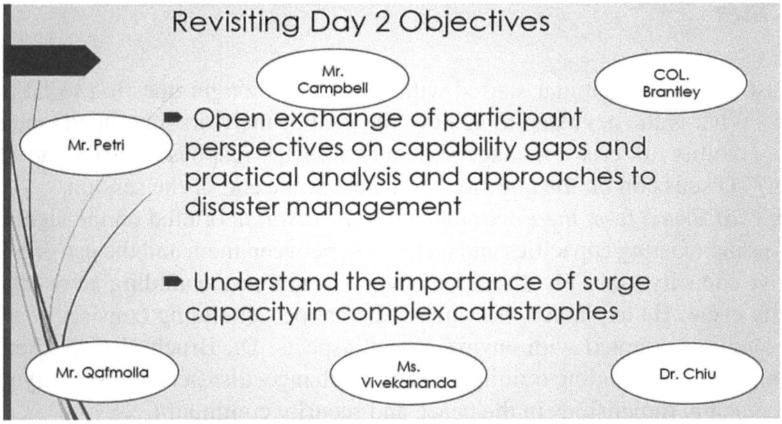

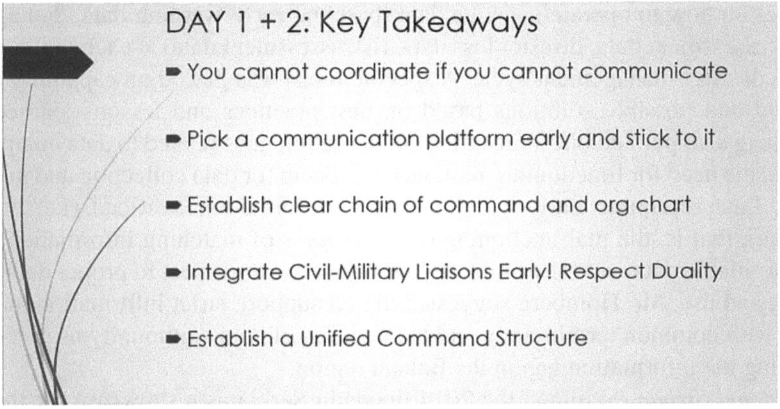

Day 3: Understanding Current Disaster Risk Management Capacity/Identifying Ways Forward

Objectives

Build mutual understanding and shared perspectives among the growing community of interests in the Balkan region. Recommend ways ahead for sustainable mechanisms for future collaboration and cooperation that include knowledge sharing and stronger multi-sector ties between the military and other stakeholders.

Activities

The last day of the seminar started with a food for thought question to all participants: "What is the next step to achieve the goal of the workshop in advancing the interoperability of crisis management and disaster response across the Balkan region?" Discussion on the questions was held at the end of the session.

Dr. Carl Bruch from the Environmental Law Institute briefed on the significance of assessing existing capacities and on the link between them and the use of building adaptive capacity tools. He discussed environmental peacebuilding across the conflict life cycle. He explained that environmental peacebuilding consists of security considerations coupled with environmental aspects. Dr. Bruch also touched upon civil-military cooperation, training on climate change, disasters and security and the environmental dimensions in the peace and security continuum.

Mr. Marc van den Homberg, representing Data for Resilience, presented opportunities on how to operate/generate data (baseline early-warning data, damage and needs assessment data, disaster loss data, risk assessment data) at each of the phases of the disaster management cycle. A specific focus was placed on capability development and possible solutions based on best practices and lessons learned. Mr. Homberg also pointed out some of the collaboration gaps related to data compatibility and the need for functioning multiagency system for data collection and management. Late receipt of early-warning data or information overload (i.e. typhoon Haiyan), that is, the malfunctioning of the process of matching information needs with available data, was one of the key challenges with respect to proper data application and use. Mr. Homberg suggested digital support, strict information management with common terminology and the creation of data community as potentially bridging the information gap in the Balkan region.

Colonel Strbac examined the 2014 floods in Serbia as a showcase for the vital importance of regional cooperation. She underlined that the international community provided aid through various international organizations, such as UN, EU, OSCE, USAID, etc., or on individual (country) bases.

Colonel Strbac remarked on the need for reviewing, revising and aligning (coordinating) prevention plans across the Balkan region and for planning and conducting joint disaster response exercises. Additionally, she identified that establishing a regional early-warning system and composing mobile teams, possessing specialized emergency response skills and equipment, would foster regional cooperation and strengthen capabilities. Colonel Strbac concluded by highlighting the main challenges to disaster response, namely, the activation of the rescue and protection teams, the insufficient rescue equipment and the complex and cumbersome coordination procedures for joint field activities.

At the beginning of the afternoon session, *Dr. Andmorgan Fisher, US Army Corps of Engineers*, outlined the topic of geoinformatics for disaster risk assessment, monitoring and management. She informed about the application of geoinformatics to the disaster management cycle. In addition, Dr. Fisher described the advantages and disadvantages of using modern data collection and imagery-gathering technologies giving the concrete examples of the Buckeye aircraft and the senseFly eBee unmanned aerial system. She further emphasized that technologies are developing rapidly which keeps costs from soaring, and, therefore, crisis management and disaster response efforts could be well completed and strengthened by modern technology.

Mr. Craig Hanrahan, Tennessee Emergency Management Agency, discussed the all-hazards approach to planning and the impact of climate change on military activities. From 2010 to 2012, the state of Tennessee weathered eight major disasters resulting in presidential declarations, including record-breaking floods in several Tennessee watersheds. Stakeholders needed innovative solutions to help address burgeoning disaster risk in the state. He also discussed how within Tennessee, following a large-scale disaster (tornadoes, floods, earthquakes, etc.), there is a need to assess buildings and structures quickly to ensure safe response and re-entry. Individuals will want to return to their homes and businesses. The Tennessee Structural Assessment and Visual Evaluation (TNSAVE) Coalition was created to help develop and maintain a post-disaster building inspector programme. This group of professional organizations consists of professional engineers, architects and building inspectors.

Lieutenant Colonel Konstantinos Brotsikas, Hellenic National Defence General Staff, brought the attention to desertification, which he pointed as one of the main challenges for the Balkan region. Lively discussion ensued after LTC Brotsikas introduced the topic of desertification reversibility. Participant pondered over the process itself and whether, if at all possible, it would be natural or human induced.

During the last day panels, Dr. Carl Bruch and Colonel Katarina Strbac delivered presentations related to theoretical and practical aspects of disaster risk management which were highly appreciated by workshop participants.

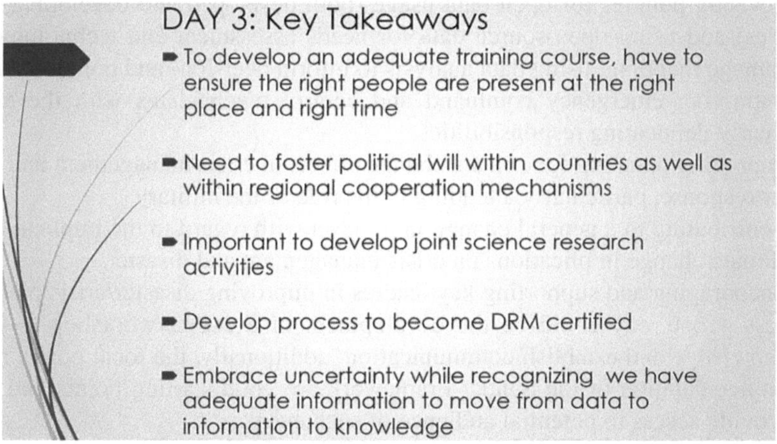

Short- and long-term actions were defined:

- Persistent networking between all participants and presenting organizations.
- More collaboration with CMDR COE.
- Build national, regional and alliance capacity.

Prevention to regional cooperation:

- Lack of commitment
- Not focused enough on planning for the future
- Lack of communication mechanism

How can we utilize this network to keep moving forward?

- Participate in CMDR COE's course.
- Tell ten people about this workshop.
- Take back lessons learned to your organization.

Impact

Based on all discussions and the aforementioned research documents, the main ARW findings and recommendations are as follows:

1. Developing an education and training event/course on the implications of climate change on crisis management and disaster response.
2. Contributing to and insisting on the inclusion of the climate change problematic on the political agenda.
3. Using tools, technology and scientific developments in support of disaster management in the context of climate change.
4. Revising policies for open data usage (both developed and developing countries) and using open-source data for needs assessment and technologies for damage mapping; using data analysis to inform decision- and policymaking.
5. Improving emergency command and control mechanisms with the aim of clearly delineating responsibilities.
6. Improving planning for climate change effects in crisis management and disaster response, particularly attending to the role of the military.
7. Contributing to a general change in mindset with regard to the implications of climate change implications on crisis management and disaster.
8. Encouraging and supporting key leaders in improving disaster/crisis preparedness – one way could be the development of a course/workshop to enrich knowledge and establish communication; additionally, the focal points for the implementation of the Sendai Framework are good starting points and could provide access to potential audience of such event.
9. Raising awareness, including through targeted strategic communication, and sharing knowledge (i.e. articles, papers and other documents) on an increase in the level of understanding and awareness of climate change and weather are

essential. The basics of weather phenomena that lead to disasters in the Balkan region as an example of such enhancement of knowledge and skills.

10. Preparing and sharing lessons learned from practice for improving disaster response and early warning.
11. Establishing a dedicated web portal (data community) for specialists, where ideas and opinions could be shared, thus contributes to regional activities and coordination.
12. Updating/improving regional agreements for waste water management.
13. Creating a regional centre of excellence with a focus on climate change.
14. Supporting the establishment of a scientific group to explore solutions in the long term.
15. Supporting and insisting on an increase in dedicated funding.
16. Including climate change in crisis management and disaster response strategies and policy frameworks.
17. Participating in a network/community of interest on climate change, i.e. providing analysis and opinion (papers, articles, blogs, etc.), and creating a clearing house to follow up on research conducted.
18. Sharing ideas, findings and recommendation to the public in order to ensure better preparedness and support.
19. Leveraging the response mechanisms of EU Civil Protection Mechanism and NATO EADRCC, USAID, as means for encouraging regional investment.
20. Fostering collaboration with the CMDR COE as a hub of expertise in the crisis management and disaster response area, i.e. creation of a science library.
21. Examining the effectiveness of initiatives undertaken at the national, regional and international levels.
22. Exploring further opportunities for enhancing regional cooperation.
23. Supporting the building of a broad regional consensus on ways to address climate change implications on crisis management and disaster response.
24. Developing regional disaster management initiatives based on successful examples from the field of security (SEEBRIG, South Eatern Europe Brigade, etc.).
25. Developing appropriate regional training and joint exercises to strengthen and foster interagency cooperation and multinational cooperation; mapping key stakeholders and their respective needs and contribution capacities.
26. Involving and engaging civil society and the private sector, particularly insurance companies, in climate change/CMDR events (i.e. video communication).

As per NATO Science for Peace and Security (SPS) programme project – SPS. ARW.985136 – predefined requirements/deliverables, an education and individual training activity (course) common core curriculum should be developed. Its main purpose is to increase climate change implication awareness. Moreover, in order to support an education and individual training activity curriculum development, a questionnaire was developed and circulated to all participants.

This questionnaire was based on NATO Bi-SC Education and Individual Training Directive (E&ITD) 075-007. Based on the above-mentioned document findings and workshop discussions, a climate change implication on military operations course common core curriculum was developed.

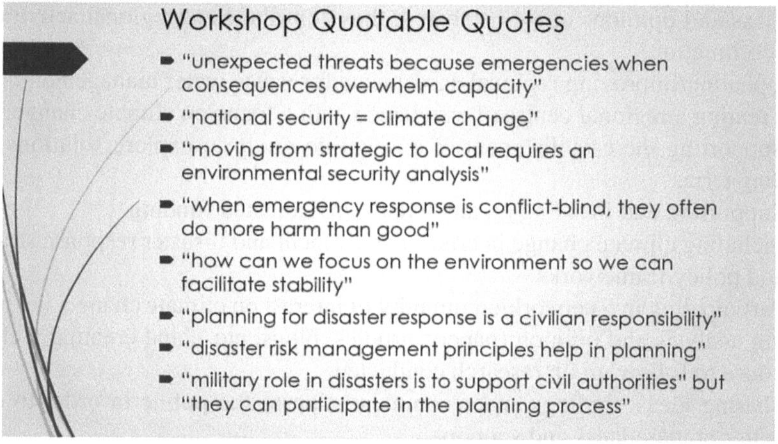

MIX
Papier aus verantwortungsvollen Quellen
Paper from responsible sources
FSC® C105338

If you have any concerns about our products,
you can contact us on
ProductSafety@springernature.com

In case Publisher is established outside the EU,
the EU authorized representative is:
Springer Nature Customer Service Center GmbH
Europaplatz 3, 69115 Heidelberg, Germany

Printed by Libri Plureos GmbH
in Hamburg, Germany